Building Pathology and Rehabilitation

Volume 24

Series Editors

Vasco Peixoto de Freitas, University of Porto, Porto, Portugal

Aníbal Costa, Aveiro, Portugal

João M. P. Q. Delgado, University of Porto, Porto, Portugal

This book series addresses the areas of building pathologies and rehabilitation of the constructed heritage, strategies, diagnostic and design methodologies, the appropriately of existing regulations for rehabilitation, energy efficiency, adaptive rehabilitation, rehabilitation technologies and analysis of case studies. The topics of Building Pathology and Rehabilitation include but are not limited to - hygrothermal behaviour - structural pathologies (e.g. stone, wood, mortar, concrete, etc…) - diagnostic techniques - costs of pathology - responsibilities, guarantees and insurance - analysis of case studies - construction code - rehabilitation technologies - architecture and rehabilitation project - materials and their suitability - building performance simulation and energy efficiency - durability and service life.

António C. Azevedo · Fernando A. N. Silva ·
João M. P. Q. Delgado · Isaque Lira

Concrete Structures Deteriorated by Delayed Ettringite Formation and Alkali-Silica Reactions

António C. Azevedo
CONSTRUCT-LFC, Faculty of Engineering
University of Porto
Porto, Portugal

Fernando A. N. Silva
Department of Civil Engineering
Universidade Catoica de Pernambuco
Recife, Pernambuco, Brazil

João M. P. Q. Delgado
CONSTRUCT-LFC, Faculty of Engineering
University of Porto
Porto, Portugal

Isaque Lira
Department of Civil Engineering
Universidade Catoica de Pernambuco
Recife, Pernambuco, Brazil

ISSN 2194-9832 ISSN 2194-9840 (electronic)
Building Pathology and Rehabilitation
ISBN 978-3-031-12269-9 ISBN 978-3-031-12267-5 (eBook)
https://doi.org/10.1007/978-3-031-12267-5

This Springer imprint is published by the registered company Springer Nature Switzerland AG
The registered company address is: Gewerbestrasse 11, 6330 Cham, Switzerland

Preface

In recent years, researches works have reported with relative frequency the occurrence of Delayed Ettringite Formation (DEF) expansions in the most varied concrete constructions, such as bridges, dams, railroads and motorways, beams and columns, spread footing and deep foundations. Most of this research state that the effects of internal expansion reactions in concrete elements might be associated with a significant change in strength and deformation properties of the material.

This book presents a discussion on the behaviour of isolated concrete bottle-shaped struts affected by internal expansion reactions. The numerical modelling of concrete was performed with the Concrete Damaged Plasticity Model (CDPM) implemented in ABAQUS and the validation of the model was performed with Sankovich's tests (Sankovich, 2003). A procedure to automatically obtain the concrete plasticity and damage parameters necessary for the use of CDPM was developed in MATLAB, based on the work of Alfarah et al. (2017), having as inputs the characteristic compressive strength of the concrete, the equivalent length of the finite element mesh and the ratio between the plastic and inelastic compressive strains. The book is divided into seven chapters.

Chapter 1 presents a characterization and the objectives of this work, namely, the justification of the research, the materials and methods used, the declaration of the general objective and the specific objectives and the limitations of the work.

Chapter 2 contemplates a literature review with a revision of the models of connecting rods and rods applied to rigid blocks of concrete on piles. This chapter also includes a discussion on bottle-shaped concrete connecting rod testing and a brief review of expansion reactions in concrete focusing on the aggregate alkali reaction.

Chapter 3 presents the fundamentals plasticity model of the damage used—Concrete Damaged Plasticity Model—CDPM—with information that allows the understanding of its potentialities and limitations. This chapter also includes the presentation of the details of the physical representation of the behaviour of concrete in the traction and compression regime and also the detailed description of the finite element characteristics used in the various numerical modelling performed in the research. The chapter also presents the complete detailing of the two routines developed in the research that serve as generators of CDPM inputs.

Chapter 4 presents in detail the experimental essays conducted by Sankovich (2003), with the detailing of the instrumentation performed, results obtained by the author and other results presented in a subsequent report of the same author with the participation of other authors.

Chapter 5 contemplates in detail the various numerical analyses performed. It includes a detailed description of the numerical models (geometry, loads and boundary conditions).

Chapter 6 presents the analysis and discussion of results. In one section it is about the validation of the model, in the next a history of the results, and in the subsequent results of the simulations with mechanical reduction of the properties of the concrete.

Finally, Chap. 7 presents the conclusions obtained in the development of this dissertation and suggestions for topics for the development of future research on the subject investigated.

Porto, Portugal — António C. Azevedo
Recife, Brazil — Fernando A. N. Silva
Porto, Portugal — João M. P. Q. Delgado
Recife, Brazil — Isaque Lira

Contents

Chapter 1
Introduction

1.1 Description

Concrete structures can be divided into regions in which the hypotheses of the theory of bending and linear distribution of deformations are applicable and others, in the vicinity of sudden changes in geometry or loading, where the distribution of deformations is no longer linear. These regions are often referred to as Regions-B and D-regions, respectively—the letter B comes from Bernoulli and the letter D comes from Discontinuity.

For B-regions, the usual bending theory for reinforced concrete elements and the traditional approach to shear (cutting resistance is broken down between concrete and steel) offer good answers, but for D-regions most of the loading is transferred to supports through flat compression forces in concrete and tensile forces in the armature and, in this way, a new design approach is needed.

Usually, the D-regions are represented with hypothetical truss models in which the connecting rods represent the compressed concrete, the rods represent the trading reinforcements, joined in points called nodes. This truss model is referred to as the Connecting Rod Model.

Concrete blocks are volume structures that have the function of distributing the loads of the pillars to deep foundation elements such as piles. In the design of this structural element it is common to use the method of connecting rods and rods and, in general, the bottle-shaped connecting rods form inside this element.

In the last decade, the occurrence of cases of early deterioration of foundation blocks on piles in residential buildings and concrete bridges in the Metropolitan Region of Recife has been reported relatively frequently. This process usually starts from the occurrence of a large horizontal fissure on the lateral faces of the element located approximately 30 cm from the upper face of the block (Silva 2007; Gomes 2008) and its installation is usually attributed to concrete expansions, due to the aggregate alkali reaction.

A. C. Azevedo et al., *Concrete Structures Deteriorated by Delayed Ettringite Formation and Alkali-Silica Reactions*, Building Pathology and Rehabilitation 24,
https://doi.org/10.1007/978-3-031-12267-5_1

In many reported cases the reinforcements located on the lower side of the block are evenly distributed, rather than being concentrated on the cutting heads, and the vertical stretch of the lower frame of the lifts has no pass with the upper mesh. Figures 1.1, 1.2 and 1.3 illustrate the pathological manifestations reported in pile crowning blocks in the region. Figure 1.3 shows the typical detail of these foundation blocks where you can see the absence of continuity of the lower and upper reinforcements of the blocks.

Regardless of the origin of the cracking process—if it comes from expansive reactions in the concrete or if it is due to the detailing of the anchorage of the vertical reinforcement of the block—its occurrence provides preferential paths for the penetration of aggressive external agents into the block that can cause deleterious effects.

The aggregate alkali reaction (AAR) is a chemical reaction in which the sodium and potassium ions contained in the solution react with certain types of rocks used in obtaining aggregates used in the manufacture of concrete and constitutes one of the important problems of early degradation of concrete structures. Water is an essential factor for the installation of the reaction, since it is necessary to form the expansive

Fig. 1.1 Large-opening horizontal fissure in pile crowning block (Silva 2007)

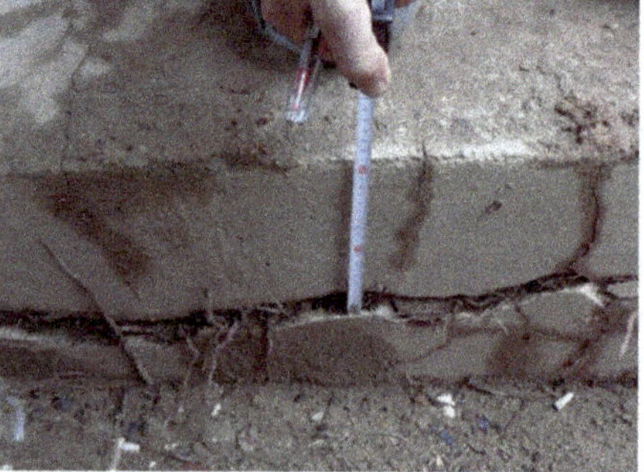

Fig. 1.2 Detailed view of horizontal fissure in pile crownblocks (Silva 2007)

Fig. 1.3 Relative horizontal displacement in foundation block (Gomes 2008)

gel that fills the pores of the concrete, resulting in undesirable expansions in the affected element. In our region, there are ideal conditions for the development of the AAR because all the elements necessary for its occurrence are present—cement alkalis or aggregates, water and large or fine aggregate potentially reactive.

The occurrence of AAR in a concrete element can generate an important reduction in its resistance and deformation properties, with more pronounced changes in its tensile strength and modulus of elasticity than in its compressive strength (Pleau et al. 1989; Reinhardt and Mielich 2102; Mielich et al. 2015; Sobrinho 2012).

This aspect becomes especially relevant when the integrity of these properties is fundamental in the overall behaviour of the affected concrete element, such as pile crowning blocks, and knowledge of the magnitude of this change constitutes a key research point applied in the field of civil engineering. In the specific case of reducing the tensile strength of concrete, its consequence on the load capacity of the pile blocks is relevant because this reduction can cause a cracking process in the connecting rods that reduces its load-sustaining capacity.

Generally, in usual field situations, the cracking process due to internal expansions does not usually lead to the rupture of the affected elements, but favours the penetration of aggressive agents such as chlorides that, reaching the reinforcements, can initiate a localized corrosion process that culminates in significantly reducing the durability of the structures.

Pathological manifestations associated with expansive internal reactions in concrete are more common in works where the volume of concrete is high such as dams and airport floors. On the other hand, in Brazil, especially in the north-east region of the country, the occurrence of the phenomenon has been reported in crown blocks of piles and shoes of residential buildings and bridges, at relatively early ages. This fact has drawn the attention of the local construction industry and research centers and universities to the need to understand the mechanisms and phenomena involved in the internal reactions of concrete expansion so that mitigating measures, both in the design and execution phase, can be taken.

Because of these events, civil engineers are ultimately having to deal with the increasing occurrence of internal expansion reactions in newly constructed concrete works without the availability of information on the influence this phenomenon exerts on the load capacity of the affected elements.

Concrete bottle-shaped connecting rods are elements that represent very efficiently the behaviour of the compression stress field inside pile crown blocks and are usually used to evaluate the level of tension and strain due to the applied loads.

Thus, the study of this configuration of connecting rods, especially when it is deteriorated by internal reactions of concrete expansion can generate important information for the understanding of the structural behaviour of pile blocks affected by the reaction that may be useful in defining design strategies and recovery of these structural elements, a topic of remarkable relevance for structural engineers and for the construction industry in the region and country.

1.2 Objectives

This book presents a discussion on the behaviour of isolated concrete bottle-shaped struts affected by internal expansion reactions. The numerical modelling of concrete was performed with the Concrete Damaged Plasticity Model (CDPM) implemented in ABAQUS and the validation of the model was performed with Sankovich's tests (Sankovich 2003). A procedure to automatically obtain the concrete plasticity and damage parameters necessary for the use of CDPM was developed in MATLAB, based on the work of Alfarah et al. (2017), having as inputs the characteristic compressive strength of the concrete, the equivalent length of the finite element mesh and the ratio between the plastic and inelastic compressive strains.

As specific objectives, the work aims to develop the following:

- Perform numerical simulations with the finite element method in linear regime in order to access the stress profile of strain stress of isolated connecting rods of simple concrete in bottle form and evaluate the various factors intervening in this behaviour;
- Perform numerical simulations with the finite element method in a nonlinear regime in order to access the stress profile of strain stress of isolated connecting rods of simple concrete in the form of bottle after cracking;
- Perform numerical modelling with the finite element method in a nonlinear regime for various levels of expansion and formulate an understanding of how expansions affect the load-sustaining mechanism of the investigated panel.

However, should be mentioned that:

- The internal expansion reactions were not explicitly modelled, but were considered through the effects that the expansions exert on the properties of resistance and deformation of concrete, according to the researches of Sanchez et al. (2014, 2015, 2016, 2017, 2018).

- The obtaining of the mechanical properties of the concrete was conducted using the standards of the ACI and CEB, the latter being very close to NBR 6118 (2014) when the guidelines for obtaining these properties.
- The damage criterion that was used by Alfarah et al. (2017) where damage is sensitive to mesh length, compression and traction fracture energies, and the relationship between plastic and inelastic compression deformation.

References

Alfarah B, López-Almansa F, Oller S (2017) New methodology for calculating damage variables evolution in plastic damage model for RC structures. Eng Struct 132:70–86

Gomes EAO (2008) Structural retrofitting works of pile caps foundations deteriorated by alkali aggregate reaction—Recife experience. MSc thesis, Catholic University of Pernambuco, Brazil

Mielich O, Reinhardt HW, Garrecht H, Giebson C, Seyfarth K, Ludwig HM (2015) Strength and deformation properties of concrete as evaluation criteria for ASR performance tests. Beton 110(2015):554–563

Pleau R, Berube MA, Pigeon M, Fournier B, Raphael S (1989) Mechanical behavior of concrete affected by ASR. In: Elsevier Applied Science (England) (eds) Proceedings of the 8th ICAAR, Kyoto, Japan, pp 721–726

Reinhardt HW, Mielich O (2012) Mechanical properties of concretes with slowly reacting alkali sensitive aggregates. In: Proceedings of the 14th ICAAR conference, Austin, TX, USA

Sanchez LFM, Multon S, Sellier A, Cyr M, Fournier B, Jolin M (2014) Comparative study of a chemo–mechanical modeling for alkali silica reaction (ASR) with experimental evidences. Constr Build Mater 72:301–315

Sanchez LFM, Fournier B, Jolin M, Bastien J (2015) Evaluation of the stiffness damage test (SDT) as a tool for assessing damage in concrete due to ASR: input parameters and variability of the test responses. Constr Build Mater 77:20–32

Sanchez LFM, Fournier B, Jolin M, Bastien J, Mitchell D (2016) Practical use of the stiffness damage test (SDT) for assessing damage in concrete infrastructure affected by alkali-silica reaction. Constr Build Mater 125:1178–1188

Sanchez LFM, Fournier B, Jolin M, Mitchell D, Bastien J (2017) Overall assessment of alkali-aggregate reaction (AAR) in concretes presenting different strengths and incorporating a wide range of reactive aggregate types and natures. Cem Concr Res 93:17–31

Sanchez LFM, Drimalas T, Fournier B, Mitchell D, Bastien J (2018) Comprehensive damage assessment in concrete affected by different internal swelling reaction (ISR) mechanisms. Cem Concr Res 107:284–303

Sankovich CL (2003) An explanation of the behavior of bottle-shaped struts using stress fields. MSc thesis, University of Texas, Austin, USA

Sobrinho CWAP (2012) Piles caps of buildings affected by AAR—case study. In: 54th Brazilian concrete congress, Maceió, Brazil (in Portuguese)

Silva GA (2007) Retrofitting works of pile caps foundations affected by alkali aggregate reaction. MSc thesis, Catholic University of Pernambuco, Brazil

Chapter 2
Brief Literature Review

2.1 Delayed Ettringite Formation (DEF)

In recent years, researches works have reported with relative frequency the occurrence of DEF expansions in the most varied concrete constructions, such as: bridges (Boenig et al. 2001; Boenig and Boenig 2000; Zhang et al. 2019), dams (Blanco et al. 2018; Karthik et al. 2016; Ayora et al. 1998; Campos et al. 2016, 2018), railroads and motorways (Sahu and Thaulow 2004; Custódio and Ribeiro 2019), beams and columns (Karthik et al. 2016, 2020), spread footing and deep foundations (Sellier and Multon 2018; Sobrinho 2012; Delgado et al. 2021). Most of this research state that the effects of internal expansion reactions in concrete elements might be associated to a significant change in strength and deformation properties of the material. Indeed, Sanchez et al. (2018) state that part of the challenge for civil engineers, in addition to knowing the level of expansion to which the structures are exposed, is to understand how concrete mechanical properties are affected by these expansions and, also, how these expansions influence the durability of the structures of concrete. The existence of internal sulfate sources is linked to the DEF expansions (Sahu and Thaulow 2004; Sanchez et al. 2018; Mehta and Monteiro 2014; Taylor 1997; Taylor et al. 2001; Diamond (1996). Escadeillas et al. (2007) reported that only internal attacks of sulfates are related to DEF, while Collepardi (2003) considers, in addition to internal attacks, external attacks caused by sources of sulfates. Collepardi (2003) additionally, reported that the presence of water is a necessary condition for the migration of sulfate ions into the concrete. The author also reported that the deposition of ettringite occurs inside the micro-cracks and its propagation is due to the expansion of crystals of ettringite. Collepardi et al. (2004) studied three parameters related to internal attacks by sulfates: (a) the sulfate content in the clinker or cement phase, (b) the curing temperature, and (c) the presence of preliminary cracks in concrete samples. The authors concluded that at curing temperatures below 80 °C there is no risk of DEF formation. Yan et al. (2004) stated that the structure of the mortar, the morphology of the ettringite and the curing humidity are relevant issues to the

A. C. Azevedo et al., *Concrete Structures Deteriorated by Delayed Ettringite Formation and Alkali-Silica Reactions*, Building Pathology and Rehabilitation 24,
https://doi.org/10.1007/978-3-031-12267-5_2

formation of the reaction, which was observed at curing temperatures above 70 °C. Several researchers reported that high temperatures during large concrete volume casting above 70 °C are potential causes of concrete expansion due to DEF (Blanco et al. 2018; Taylor 1997; Taylor et al. 2001). Mielich et al. (2015) studied the influence of the NaCl solution as an external source of alkali on the creep behaviour of concrete. The authors reported an expansion greater than 0.5 mm/m after six cycles. The modulus of elasticity after 252 days showed a decrease of 17%. Reinhardt and Mielich (2012) reported that internal expansions in concrete are dependent on the type of coarse aggregate used. Regarding tensile strength, it was observed a decrease close to 75%. Sobrinho (2012) performed lab tests in concrete cores collected from pile caps affected by internal expansion reactions and reported a maximum decrease of 34.7% in the split tensile strength and 23% in the elastic modulus of concrete. Sanchez et al. (2017, 2018) discussed the evolution of damage caused by internal expansion reactions using multilevel approach. The authors used aggregates of different origins to produce concrete mixes, which were designed for the characteristic strength of 35 MPa. In the damage caused by DEF expansions, the authors observed maximum decrease in values of the modulus of elasticity of 86% and in the compressive strength of 50%. DEF chemical modelling for assessment of affected concrete structures was investigated by Sellier and Multon (2018). The model was fitted with different materials, heating cycles and storage conditions. The results obtained describe well the swelling potential obtained in several lab tests. The authors also showed that the output of the chemical model can be used in the poromechanical model through the concept of pressure induced by DEF in concrete. Giannini et al. (2013, 2018) reported the relevance of the stiffness damage test (SDT) to provide a wide range of deterioration levels in concretes elements affected by DEF expansion. The authors concluded that the stiffness damage index (SDI) and the plastic deformation index (PDI) proposed by Sanchez et al. (2015, 2016, 2017) can be satisfactorily used to characterize the DEF. Results of decreases in mechanical properties of concrete of 75% for its young's modulus and of 41% for its compressive strength were also reported by Giannini et al. (2013).

To perform the numerical analyses in the research, it was adopted the strategy to take into account the effects of DEF expansions in concrete through the decrease of mechanical properties of strength and deformation of the material. This approach is consistent with several studies of the literature that observed this fact in numerical modelling and laboratory tests (Campos et al. 2018; Karthik et al. 2020; Sellier and Multon 2018; Sanchez et al. 2018; Giannini et al. 2018; Seignol et al. 2009; Thiebau et al. 2018).

In resume, this work presents a numerical analysis of the structural performance of isolated concrete bottle-shaped struts deteriorate by internal expansion due to delayed ettringite formation using the finite element method. A better comprehension of the effects of DEF expansions is an important issue for engineers because they can damage concrete structures severely. By the other hand, since experimental tests are generally costly and require a long period to obtain results, the use of computer simulations may be a viable strategy if one has a reliable modelling strategy. Concrete Damaged Plasticity (CDP) model was used to perform non-linear finite elements analyses. CDP is a continuum model, plasticity-based, damage model for concrete that assumes that the main two failure mechanisms are tensile cracking and compressive crushing of the concrete material. Among the various possibilities of a phenomenological approach of DEF expansions concrete, it was adopted the strategy of taking into account the effect that those expansions generate on the mechanical properties of concrete. This modelling strategy is convergent with several studies that have already confirmed the effect of reducing the mechanical properties of concrete, in an important way, when the material is affected by internal expansion reactions due to DEF (Karthik et al. 2020; Sanchez et al. 2018; Giannini et al. 2018).

2.2 Alkali-Silica Reaction (ASR)

Nearby of sudden changes in geometry or loading, the strain distribution profile doesn't follow a linear evolution law. These regions are represented by hypothetical truss models in which the connecting struts represent the compressed concrete and the ties represent the tensile reinforcements, joined at points called nodes.

Concrete pile cap are volume structures that have the function of distributing the loads of the columns to elements of deep foundations, such as piles and piers. In the design of this structural element it is common to use the struts and ties method and, in general, bottle-shaped struts are formed inside this element.

Cases of early deterioration of foundation blocks in residential buildings and concrete bridges in the Metropolitan Region of Recife have been reported with relative frequency in the last decade. This process usually starts with the occurrence of a horizontal crack with a wide opening on the lateral faces of the element located approximately 30 cm from the upper face of the block (Silva 2007, Gomes 2008). This occurrence is usually attributed to concrete expansions, resulting from Internal Expansion Reactions (IER), most frequently the Alkali-Silica Reaction (ASR).

Internal expansion reactions are important issues in the early degradation of concrete structures. For the development of these reactions, cement alkali content, reactivity of coarse aggregates and/or the existence of a source of moisture are necessary and sufficient conditions: (Bangert et al. 2004; Lindgård et al. 2012).

The occurrence of ASR in a concrete element can generate an important reduction in its strength and deformation properties, with more pronounced effects in its tensile strength and modulus of elasticity than in its compressive strength (Hobbs 1988; Pleau et al. 1989; Reinhardt and Mielich 2012; Mielich et al. 2015; Sobrinho 2012; Sanchez et al. 2014; Diab et al. 2020; Jiang et al. 2020). In the specific case of the reduction of the tensile strength of the concrete, its consequence in the load capacity of the pile cap blocks is relevant because this reduction can cause a cracking process in the struts that reduces the element load capacity.

It is of great interest to the academic and technical community to better understand the alkali-silica reaction (ASR) because the damage it imposes to the concrete structures such as pile cap blocks, dams bridges, concrete walls, pavements and nuclear power plant usually demands a large amount of money to perform retrofitting works.

ASR is one of the main problems for durability of concrete in its occurrence can cause damages like expansion and cracking that may lead to a decrease in its strength and in its elasticity modulus (Rajabipour et al. 2015). The damage generated by ASR also affects the serviceability of concrete infrastructures (Balachandran et al. 2017).

ASR is a reaction that demands special conditions to develop, such as enough amount of alkali in pore solution (pH: 13.0–13.5) and portlandite from cement paste, certain siliceous phases in some natural and synthetic aggregates and enough moisture (greater than 70%). These combined factors results in secondary hydration products that can be lead to expansion (Rajabipour et al. 2015; Balachandran et al. 2017; Bourdot et al. 2016; Dähn et al. 2016). The alkali content of the cement is expressed as sodium oxide soluble in equivalent acid corresponding to Na_2O^+ 0,658K_2O. The alkalis in the cement usually range from 0.2 to 1.5% of Na_2O equivalent and come from the raw materials of clinker. Depending on this alkali content, the solution of the concrete pores may exhibit a pH between 12.5 and 13.5 (Merz and Leemann 2013).

The chemical process of the ASR reaction starts when the silica structure is dissolved by the nucleophilic attack of hydroxide ion (OH^-), which due to its highly degraded 2010structure, behaves like a hygroscopic silica gel. The reaction product is a gel known as Alkali Silica Gel, which is not injurious for the concrete (Ghanem et al. 2010). However, this gel has a tendency to swell when it absorbs moisture present in the concrete pore solution and, if confined in the matrix, can generate internal stresses. The more it absorbs moisture the more internal pressures increase. This phenomenon induces the development of micro-cracks, which in extreme situations, can lead to concrete failure (Rajabipour et al. 2015; Ghanem et al. 2010; Benmore and Monteiro 2010).

Chemical reactions that govern the ASR damages (Rajabipour et al. 2015) are the dissolution of metastable silica, the formation of nano-colloidal silica sol, the gelation of the sol and the swelling of the gel. Gel expansion depends on both the dissolution of reactive or potentially reactive silica and the mass transport properties of the concrete and the availability of moisture. This expansion causes less damage to dry and dense concrete. In a confined environment, concrete exhibits considerable sensitivity due to the presence of internal defects, faults and micro-cracks. Thus, solid concrete under tri-axial compression states present localized tensions because of their heterogeneous features. In view of these aspects, their behaviour becomes even more complex when damage occurs (Sanchez et al. 2018).

The pattern characteristics and extension of the cracking process provide information on the mechanism of internal expansion reactions and the magnitude of the microscopic damage to the concrete structure (Sanchez et al. 2018).

The alkali-silica reaction does not develop rapidly, taking many years for its full propagation and it is deleterious when alkalis are found in high concentration in the pore fluid (MERL 2009). The damages due to the reaction can be identified by some features exhibited in the concrete structure, such as, cracking, misalignment of structural elements, spalling and discoloration at the concrete surface, disintegration of the cement paste, surface pop-outs and gel exudation (Hobbs 1988; Rajabipour et al. 2015; Balachandran et al. 2016; Molin 1988; Thomas et al. 2011; Swamy 1992).

2.2.1 Numerical Modelling of ASR

The numerical modelling of the silica alkali reaction in concrete is not a task of simple implementation due to the complexity of the phenomena involved. In general, this modelling is usually conducted from an approach that involves the formation and expansion of kinetic chemical reactions or diffusion or a mechanical fracture approach that comes from the expansion and deterioration of the structure (Murdoch 2015).

Over the past few years, various numerical models of representation of ASR in concrete have been developed (Pignatelli 2012; Bazant and Steffens 2000; Bazant et al. 2000; Ulm et al. 2000; Steffens and Coussy 2003; Multon and Toutlemonde 2010; Multon et al. 2009; Comi et al. 2009; Grimal et al. 2008a, 2008b; Dunant 2009; Dunant and Scrivener 2010; Alnaggar et al. 2017). Bazant's work (Bazant and Steffens 2000; Bazant et al. 2000) describes the formulation of a model at the mesoscopic level for simulation of the kinetics of chemical reactions and the processes related to diffusion and fracture mechanics of the damage process. Ulm et al. (2000) formulated a chemical–mechanical model in which chemical reactions have a thermo-activated mechanism that accelerates reactions with increasing temperature. In this model, however, the authors did not include the effects of moisture (water) which plays an important role in the kinetics of reactions. In this sense, Steffens and Coussy (2003) created a model with a similar global approach, but incorporating the effects of dependence on the moisture history of alkali-silica gel. It is a model composed of two stages: (a) the formation of the amorphous gel that is dependent on moisture and (b) an instantaneous process of gel-water interaction, which is influenced by the aging of the gel caused by the action of water. Multon and Toutlemonde (2010) studied the effect of the path travelled by water on the expansion of the ASR reaction when changes in the humidity of the medium are present. The authors concluded that if at any time of the life of a structure affected by ASR, if there is a source of moisture available, the gel already produced can expand and, even in cases where the reaction has already been stopped due to the lack of water, the water source can cause a new formation of ASR gel.

Multon et al. (2009) created a microscopic model to predict the development of ASR expansion in cement mortars with aggregates of different sizes. The model is based on the theory of damage and is able to show the decreasing of mortar stiffness due to the cracking process caused by ASR. The proposed model was able to satisfactorily reproduce the dependence on expansion, on both the size of the aggregate and the alkali content of the medium.

Comi et al. (2009) proposed a thermochemical damage model to simulate the process of expansion and deterioration of concrete strength and stiffness due to alkalis-aggregate reactions. The model assumes that the concrete affected by the reaction behaves like a heterogeneous biphasic material—an expansive gel and a homogenized skeleton of concrete. The microcracking process due to gel expansion is represented by an isotropic damage approach with scalar damage variables, both in tension and compression regimes.

Bangert et al. (2004) created a macroscopic model to describe the process of deterioration of concrete caused by ASR in which the material is conceived as a three-phase mixture: unreacted material, unexpanded material, but already reacted, and expanded material. In this model, the dependence of kinetic reactions and the magnitude of expansion as a function of humidity are taken into account. The authors concluded that the deterioration of concrete imposed by ASR reactions is governed by the simultaneous action of moisture diffusion and kinetic reactions, which led to an important reduction in load capacity and stiffness of the concrete elements investigated.

Grimal et al. (2008a) presented a discussion about the elementary physical principles of an orthotropic viscoelastic-plastic damage model that includes a chemical pressure induced by alkalis-aggregate reactions. The authors also considered the modelling of the effects of moisture on the development of reactions and drying shrinkage. Grimal et al. (2008b), based on the work of Grimal et al. (2008a), calibrated the parameters of the model using analyses of reinforced concrete beams degraded by internal expansive reactions under mechanical loading and non-homogeneous moisture conditions. The model was able to reproduce the markedly anisotropic expansion, the damage observed during the experimental tests and the displacements measured.

Dunant (2009) developed a model to simulate ASR microscopically using the finite element method with a new strategy to calculate the associated damage. The model takes into account the interaction between aggregates and also the effects of transformation of the expansive gel at the individual level of the aggregates. The author reports that it is possible to model the ASR using the finite element method with few parameters to be adjusted. It also reports that the mechanical consequences of the reaction depend on the mechanical properties of ASR gel, which, in turn, are dependent on the availability of Ca^{2+} and K^{+} ions from the porous solution. As a result of this fact, the author states that the mechanical effects of the reaction are also dependent on the type of aggregate and the type of cement used to produce the concrete.

Dunant and Scrivener (2010) proposed a physical model of representation of ASR that takes into account the effect of gel formation on aggregates on the mechanisms of degradation of the reaction, mainly the damage to the aggregates themselves. According to the authors, the model was able to adequately simulate the mechanisms involved in the reaction. Macroscopic free expansion and degradation of the mechanical properties of concrete were reported to be related to the extent of the reaction.

Alnaggar et al. (2017) proposed a model that couples creep, shrinkage and swelling due to ASR using the Lattice Discrete Particle Model (LDPM). A multi-physical formulation was used to calculate the evolution of temperature, moisture, cement hydration and ASR in space and time. The model was calibrated with experiments available in the literature and the results obtained showed that, even during free expansion, an important degree of coupling exists because the ASR induced expansions were relieved by the mesoscale creep.

In this work it was decided to consider the effects of internal expansion reactions in concrete by adopting a strategy that its occurrence in a concrete element leads to a decrease of the its mechanical properties of strength and deformation. This approach is consistent with several studies that observed this fact in numerical modelling and laboratory tests Sanchez et al. 2014, 2017, 2018; Pignatelli 2012; Bazant and Steffens 2000; Bazant et al. 2000; Ulm et al. 2000; Steffens and Coussy 2003; Multon and Toutlemonde 2010; Multon et al. 2009; Comi et al. 2009; Grimal et al. 2008a, 2008b; Dunant 2009; Dunant and Scrivener 2010; Alnaggar et al. 2017).

References

ABAQUS (2018) ABAQUS user's manual. Dassault Systèmes, Providence, Rohde Island, USA. Simulia Corporation ABAQUS vs. 6.18

Alfarah B, López-Almansa F, Oller S (2017) New methodology for calculating damage variables evolution in plastic damage model for RC structures. Eng Struct 132:70–86

Alnaggar M, Luzio GD, Cusatis G (2017) Modeling time-dependent behavior of concrete affected by alkali silica reaction in variable environmental conditions. Materials 10(5):471

Ayora C, Chinchón S, Aguado A, Guirado F (1998) Weathering of iron sulfides and concrete alteration: thermodynamic model and observation in dams from central pyrenees, Spain. Cem Concr Res 28(9):1223–1235

Balachandran C, Muñoz JF, Arnold T (2017) Characterization of alkali silica reaction gels using Raman spectroscopy. Cem Concr Res 92:66–74

Bangert F, Kuhl D, Meschke G (2004) Chemo-hygro-mechanical modelling and numerical simulation of concrete deterioration caused by alkali-silica reaction. Int J Numer Anal Methods Geomech 28(78):689–714

Bazant ZP, Steffens A (2000) Mathematical model for kinetics of alkali–silica reaction in concrete. Cem Concr Res 30:419–428

Bazant ZP, Zi G, Mayer C (2000) Fracture mechanics of ASR in concretes with waste glass particles of different sizes. J Eng Mech 126(3):226–232

Benmore CJ, Monteiro PJM (2010) The structure of alkali silicate gel by total scattering methods. Cem Concr Res 40(6):892–897

Blanco A, Cavalaro SHP, Segura I, Segura-Castillo L, Aguado A (2018) Expansions with different origins in a concrete dam with bridge over spillway. Constr Build Mater 163:861–874

Boenig A, Boenig A (2000) Bridges with premature concrete deterioration: field observations and large-scale testing. MSc thesis, University of Texas at Austin, USA

Boenig A, Funez L, Klingner RE, Fowler TJ (2001) Bridges with premature concrete deterioration: field observations and large-scale testing, N. FHWA/TX-02/1857–1, Center for Transportation Research, University of Texas at Austin, USA

Bourdot A, Thiéry V, Bulteel D, Hammerschlag JG (2016) Effect of burnt oil shale on ASR expansions: a petrographic study of concretes based on reactive aggregates. Constr Build Mater 112:556–569

Brown MD, Sankovich CL, Bayrak O, Jirsa JO, Breen JE, Wood SL (2006a) Design for shear in reinforced concrete using strut-and-tie models. Report No. FHWA/TX-06/0-4371-2, University of Texas at Austin

Brown D, Sankovich CL, Bayrak O, Jirsa JO (2006b) Behavior and efficiency of bottle shaped struts. ACI Struct J 103(3):348–355

Campos A, Lopez CM, Blanco A, Aguado A (2016) Structural diagnosis of a concrete dam with cracking and high nonrecoverable displacements. J Perform Constr Fac 30(5):04016021

Campos A, Lopez CM, Blanco A, Aguado A (2018) Effects of an internal sulfate attack and an alkali-aggregate reaction in a concrete dam. Constr Build Mater 166:668–683

Collepardi M (2003) A state-of-art review on delayed ettringite attack on concrete. Cem Concr Compos 25:401–407

Collepardi M, Olagot JO, Salvioni D, Sorrentino D (2004) DEF-related expansion of concrete as a function of sulfate content in the clinker phase or cement and curing temperature. Sympos Paper 222:77–92

Comi C, Fedele R, Perego U (2009) A chemo-thermo-damage model for the analysis of concrete dams affected by alkali–silica reaction. Mech Mater 41(3):210–230

Custódio J, Ribeiro AB (2019) Evaluation of damage in concrete from structures affected by internal swelling reactions–a case study. Procedia Struct Integrity 17:80–89

Dähn R, Arakcheeva A, Schaub Ph, Pattison P, Chapuis G, Grolimund D, Wieland E, Leemann A (2016) Application of micro X-ray diffraction to investigate the reaction products formed by the alkali–silica reaction in concrete structures. Cem Concr Res 79:49–56

Delgado JMPQ, Nascimento N, Silva FAN, Azevedo AC (2021) Diagnostic of concrete samples of pile caps affected by internal swelling reactions, Iran. J Sci Technol Trans Civ Eng

Diab SH, Soliman AM, Nokken MR (2020) Changes in mechanical properties and durability indices of concrete undergoing ASR expansion. Constr Build Mater 251:118951

Diamond S (1996) Delayed ettringite formation-processes and problems. Cem Concr Compos 18(3):205–215

Dunant C (2009) Experimental and modelling study of the alkali-silica-reaction in concrete. PhD Thesis, Lausanne, EPFL, Switzerland

Dunant CF, Scrivener KL (2010) Micro-mechanical modelling of alkali-silica–reaction-induced degradation using the AMIE framework. Cem Concr Res 40(4):517–525

Escadeillas G, Aubert JE, Segerer M, Prince W (2007) Some factors affecting delayed ettringite formation in heat-cured mortars. Cem Concr Res 37(10):1445–1452

Ghanem H, Zollinger D, Lytton R (2010) Predicting ASR aggregate reactivity in terms of its activation energy. Constr Build Mater 24(7):1101–1108

Giannini ER, Folliard KJ, Zhu J, Bayrak O, Kreitman K, Webb Z, Hanson B (2013) Non-destructive evaluation of in-service concrete structures affected by alkali-silica reaction (ASR) or delayed ettringite formation (DEF). Final report, part I, No. FHWA/TX-13/0–6491–1, Center for Transportation Research, Austin, Texas, USA

Giannini ER, Sanchez LFM, Tuinukuafe A, Folliard KJ (2018) Characterization of concrete affected by delayed ettringite formation using the stiffness damage test. Constr Build Mater 162:253–264

Gomes EAO (2008) Structural retrofitting works of pile caps foundations deteriorated by alkali aggregate reaction–Recife experience. MSc Thesis, Catholic University of Pernambuco, Brazil

Grimal E, Sellier A, Le Pape Y, Bourdarot E (2008a) Creep, shrinkage, and anisotropic damage in AAR swelling mechanism–Part I: A constitutive model. ACI Mater J 105(3):227–235

Grimal E, Sellier A, Le Pape Y, Bourdarot E (2008b) Creep, shrinkage and anisotropic damage in AAR swelling mechanism–Part II: Identification of model parameters and application. ACI Mater J 105(3):236–242

Hobbs DW (1988) Alkali–silica Reaction in Concrete. Thomas Telford, London, UK

Jiang Z, He B, Zhu X, Ren Q, Zhang Y (2020) State-of-the-art review on properties evolution and deterioration mechanism of concrete at cryogenic temperature. Constr Build Mater 257:119456

Karthik MM, Mander JB, Hurlebaus S (2016) Deterioration data of a large-scale reinforced concrete specimen with severe ASR/DEF deterioration. Constr Build Mater 124:20–30

Karthik MM, Mander JB, Hurlebaus S (2020) Simulating behaviour of large reinforced concrete beam-column joints subject to ASR/DEF deterioration and influence of corrosion. Eng Struct 222:111064

Laughery L, Pujol S (2015) Compressive strength of unreinforced struts. ACI Struct J 112(5):617–624

Lindgård J, Andiç-Çakir Ö, Fernandes I, Rønning TF, Thomas MDA (2012) Alkali–silica reactions (ASR): Literature review on parameters influencing laboratory performance testing. Cem Concr Res 42(2):223–243

Mehta PK, Monteiro PJM (2014) Concrete: microstructure, properties e materials, 4th edn. McGraw-Hill Professional Publishing, USA

MERL Report-09-23 (2009) New recommendations for ASR mitigation in reclamation concrete construction. U.S. Bureau of Reclamation, Denver, Colorado, USA

Merz C, Leemann A (2013) Assessment of the residual expansion potential of concrete from structures damaged by AAR. Cem Concr Res 52:182–189

Mielich O, Reinhardt HW, Garrecht H, Giebson C, Seyfarth K, Ludwig HM (2015) Strength and deformation properties of concrete as evaluation criteria for ASR performance tests. Beton 110(2015):554–563

Multon S, Sellier A, Martin C (2009) Chemo-mechanical modeling for predition of alkali silica reaction (ASR) expansion. Cem Concr Res 39:490–500

Multon S, Toutlemonde F (2010) Effect of moisture conditions and transfers on alkali silica reaction damaged structures. Cem Concr Res 40:924–934

Murdoch R (2015) Creating a simplified model of the alkali silica reaction in concrete by utilizing finite element modelling techniques. MSc Thesis, University of Southern Queensland, Australia

Pignatelli R (2012) Modeling of degradation induced by alkali-silica reaction in concrete structures. PhD Thesis, POLIMI, Italy

Pleau R, Berube MA, Pigeon M, Fournier B, Raphael S (1989) Mechanical behavior of concrete affected by ASR, In: Elsevier Applied Science (England) (Eds.), Proceedings of the 8th ICAAR, Kyoto, Japan, 721–726

Rajabipour F, Giannini E, Dunant C, Ideker JH, Thomas MD (2015) Alkali–silica reaction: current understanding of the reaction mechanisms and the knowledge gaps. Cem Concr Res 76:130–146

Reinhardt HW, Mielich O (2012) Mechanical properties of concretes with slowly reacting alkali sensitive aggregates. In: Proceedings of the 14th ICAAR conference, Austin, TX, USA

Rossi PP (2013) Evaluation of the ultimate strength of RC rectangular columns subjected to axial force, bending moment and shear force. Eng Struct 57:339–355

Sahoo DK, Gautam RK, Singh B, Bhargava P (2008) Strength and deformation of bottle-shaped struts. Magaz Concr Res 60(2):137–144

Sahu S, Thaulow N (2004) Delayed ettringite formation in Swedish concrete railroad ties. Cem Concr Res 34:1675–1681

Sanchez LFM, Drimalas T, Fournier B, Mitchell D, Bastien J (2018) Comprehensive damage assessment in concrete affected by different internal swelling reaction (ISR) mechanisms. Cem Concr Res 107:284–303

Sanchez LFM, Fournier B, Jolin M, Bastien J (2015) Evaluation of the Stiffness Damage Test (SDT) as a tool for assessing damage in concrete due to ASR: input parameters and variability of the test responses. Constr Build Mater 77:20–32

Sanchez LFM, Fournier B, Jolin M, Bastien J, Mitchell D (2016) Practical use of the Stiffness Damage Test (SDT) for assessing damage in concrete infrastructure affected by alkali-silica reaction. Constr Build Mater 125:1178–1188

Sanchez LFM, Fournier B, Jolin M, Mitchell D, Bastien J (2017) Overall assessment of Alkali-Aggregate Reaction (AAR) in concretes presenting different strengths and incorporating a wide range of reactive aggregate types and natures. Cem Concr Res 93:17–31

Sanchez LFM, Multon S, Sellier A, Cyr M, Fournier B, Jolin M (2014) Comparative study of a chemo–mechanical modeling for alkali silica reaction (ASR) with experimental evidences. Constr Build Mater 72:301–315

Sankovich CL (2003) An explanation of the behavior of bottle-shaped struts using stress fields. MSc thesis, University of Texas, Austin, USA

Seignol JF, Baghdadi N, Toutlemond F (2009) A macroscopic chemo-mechanical model aimed at re-assessment of delayed ettringite formation affected concrete structures. In: 1st international conference on computational technologies in concrete structures CTCS'09, Jeju, Korea

Sellier A, Multon S (2018) Chemical modelling of delayed ettringite formation for assessment of affected concrete structures. Cem Concr Res 108:72–86

Silva GA (2007) Retrofitting works of pile caps foundations affected by alkali aggregate reaction. MSc Thesis, Catholic University of Pernambuco, Brazil

Sobrinho CWAP (2012) Piles caps of buildings affected by AAR—case study. In: 54th Brazilian concrete congress, Maceió, Brazil (in Portuguese)

Steffens A, Li K, Coussy O (2003) Ageing approach to water effect on alkali-silica reaction. Degradation of structures. J Eng Mech 129:50–59

Swamy RN (1992) The alkali-silica reaction in concrete. Taylor-Francis, CRC Press, USA

Taylor HF, Famy C, Scrivener K (2001) Delayed ettringite formation. Cem Concr Res 31(5):683–693

Taylor HFW (1997) Cement chemistry, 2nd edn. Thomas Telford Publishing, UK, London

Thiebaut Y, Multon S, Sellier A et al (2018) Effects of stress on concrete expansion due to delayed ettringite formation. Constr Build Mater 183:626–641

Thomas MDA, Fournier B, Folliard KJ, Resendez YA (2011) Alkali-silica reactivity field identification handbook, No. FHWA-HIF-12-022, US Department of Transportation, USA

Ulm FJ, Coussy O, Kefei L, Larive C (2000) Thermo-chemo-mechanics of ASR expansion in concrete structures. J Eng Mech 126:233–242

Yan P, Zheng F, Peng J, Qin X (2004) Relationship between delayed ettringite formation and delayed expansion in massive shrinkage-compensating concrete. Cem Concr Compos 26(6):687–693

Zhang H, Li L, Wang W (2019) Effects of temperature rising inhibitor on nucleation and growth process of ettringite. J Solid State Chem 274:222–228

Chapter 3
The Plasticity Model of Concrete Damage—CDPM

Numerical simulation of the behaviour of concrete structures is not a task of simple understanding and implementation. Various phenomena related to the rheology of the material itself associated with the interaction of concrete with steel (reinforced concrete) contribute to lend a particular complexity that are not frequent in other construction materials. Among these phenomena deserve to be highlighted: the micro-cracking of concrete that occurs in the early ages as a consequence of non-conformities in its healing process, the cracking in traction regime of the material that originates from the loadings imposed on it, heterogeneity and anisotropy of the concrete constituents, the phenomena involved in steel–concrete adhesion; the rheological behaviour of concrete that generates changes in its resistance and deformation properties over time.

In view of the aspects mentioned above, the models of computational representation of concrete behaviour usually involve in its formulation varied approaches as a strategy to capture the influence of most of the phenomena reported. In the practice of advanced research of the response of concrete structures submitted to various loading conditions using computational numerical analyses, there is a relative availability of software, mostly based on the Finite Element Method (FEM), and its widespread use is widely disseminated as follows: ABAQUS, ADYNA, ANSYS, COMSOL, DIANA, NASTRAN and PATRAN.

The numerical modeling carried out throughout the development of the dissertation used the model based on the damage approach, developed based on the plasticity theory, implemented in the ABAQUS program (Hibbit et al. 2012) under the name concrete damaged plasticity model (CDPM). It is a robust model of physical representation of concrete behavior, under traction and compression regime, which incorporates the concept of isotropic elastic damage to describe the nonlinear behavior of concrete. It was initially designed by Lubliner et al. (1989) and later modified by Lee and Fenves (1998).

A. C. Azevedo et al., *Concrete Structures Deteriorated by Delayed Ettringite Formation and Alkali-Silica Reactions*, Building Pathology and Rehabilitation 24,
https://doi.org/10.1007/978-3-031-12267-5_3

The definition of a numerical model based on the Plasticity Theory requires the explanation of a flow criterion, the formulation of a constitutive law of hardening/softening and the establishment of a flow law. Each of these topics will be covered in the following sections because these concepts will serve as the basis for future discussions about the ABAQUS CDP Model.

3.1 Additive Decomposition of the Field of Total Deformations

The additive decomposition of total deformation is a foundation of the models based on the Plasticity Theory. It expresses itself from Eq. (3.1) as:

$$\varepsilon = \varepsilon^{el} + \varepsilon^{pl} \tag{3.1}$$

where ε is the total deformation that is decomposed into two plots: an elastic—ε^{el}—and a plastic—ε^{pl}, usually referred to as *Plastic Strain* (PE).

Similarly the additive decomposition of the total deformation of application of the model, the deformation rate consisting of the consideration of the gradient of the deformation velocity, is calculated according to Eq. (3.2), as follows:

$$\varepsilon^{\cdot} = \varepsilon^{\cdot el} + \varepsilon^{\cdot pl} \tag{3.2}$$

where $\varepsilon^{\cdot}$ is the rate of total deformations, $\varepsilon^{\cdot el}$ is the rate of elastic deformations and $\varepsilon^{\cdot pl}$ is the rate of plastic deformations.

The nonlinear behaviour of the concrete is based on the concept of isotropic elastic damage as a strategy for representing the degradation of material stiffness in conjunction with irreversible damage, which manifests itself during the material rupture process, in combination with isotropic plasticity to describe the mechanisms of damage—i.e., softening in traction regime and crushing under compression regime.

Local damage models assume that nonlinear behaviour is controlled by two different mechanisms, one in traction and one in compression. To establish the elasto-plastic constitutive equations of the model, an initial flow surface, a hardening law, a no associative flow law and a Drucker-Prager hyperbolic plastic potential function are used.

3.2 Mechanical Behaviour of the Model

The model admits that the main mechanisms of concrete breakage are traction cracking and compression crushing. The evolution of the rupture surface is controlled by two hardening variables $\tilde{\varepsilon}_t^{pl}$ and $\tilde{\varepsilon}_c^{pl}$, associated with tensile and compression rupture mechanisms, respectively, and are usually referred to as equivalent plastic deformations in traction and compression.

3.2.1 Uniaxial Behaviour in Traction and Compression

The model assumes that the response of concrete in traction and axial compression is characterized by the plasticity of the damage, as illustrated in Fig. 3.1.

As shown in Fig. 3.1b, the response of the concrete in uniaxial tensile, in terms of the stress–strain relationship, is initially linear until the actuating stress reaches the rupture limit of the material in traction—σ_{t0}. The rupture stress corresponds to the beginning of the formation of the micro-cracking process in the concrete material. In addition to this rupture stress value, the formation of micro-cracks is represented macroscopically through the phenomenon of softening in the stress–strain response of the material, which induces localized deformation in the concrete structure. The CDP model includes in its constitutive functions the visco-plastic regularization, to take into account the localized deformation, through control of deformation states of the points.

In uniaxial compression regime—Fig. 3.1a—the material response is linear until the stress value corresponding to the yield limit of the material—σ_{c0}. In the plastic stretch, the response is characterized by a hardening behaviour followed by a softening behaviour, after the rupture stress of the compressed material—σ_{cu}. These phenomenological representations of the behaviour of concrete in uniaxial traction and compression, despite being relatively simplified, capture very efficiently the response of the material when submitted to the two states of tension discussed. In the model formulation, uniaxial stress–strain curves can be converted into plastic stress-deformation curves. This conversion is performed automatically in ABAQUS from the user-provided inelastic stress–strain data, using Eqs. (3.3) and (3.4),

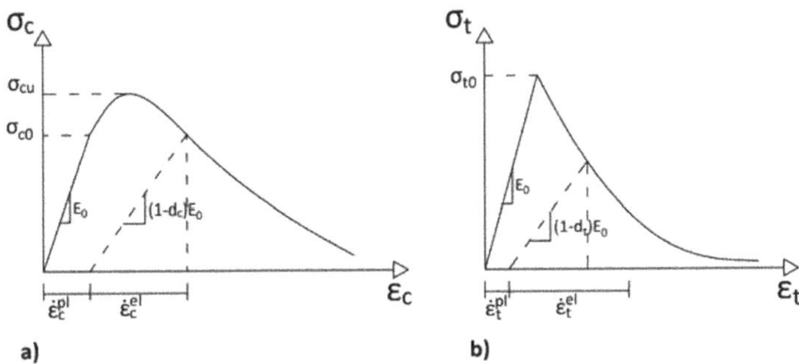

Fig. 3.1 Uniaxial response of concrete **a** in compression **b** in traction (Hibbit et al. 2012)

$$\sigma_t = \sigma_t(\varepsilon_t^{\sim pl}, \varepsilon_t^{*pl}, \theta, f_i) \tag{3.3}$$

$$\sigma_c = \sigma_c\left(\varepsilon_c^{\sim pl}, \varepsilon_c^{*pl}, \theta, f_i\right) \tag{3.4}$$

where the subscripts t and c refer to traction and compression, respectively; $\varepsilon_t^{\sim pl}$ and $\varepsilon_c^{\sim pl}$ are equivalent plastic deformations; ε_t^{*pl} and ε_c^{*pl} are the rates of variation of equivalent plastic deformations; θ is the temperature and f_i ($i = 1, 2, \ldots$) are other predefined field variables.

As can be seen in Fig. 3.1, since the concrete is discharged from some point in the softening portion of the stress x deformation curve, the discharging response of the material is weakened and its elastic stiffness degrades. The degradation of the elastic stiffness of the concrete is introduced, in turn, using two damage parameters that are dependent on plastic deformations, temperature and other field variables, according to Eqs. (3.5) and (3.6)

$$d_t = d_t\left(\varepsilon_t^{\sim pl}, \theta, f_i\right); \quad 0 \le d_t \le 1 \tag{3.5}$$

$$d_c = d_c\left(\varepsilon_c^{\sim pl}, \theta, f_i\right); \quad 0 \le d_c \le 1 \tag{3.6}$$

3.2.2 Nonlinear Behaviour in Traction

The nonlinear behaviour in traction is represented in Fig. 3.2. The nonlinear stretch begins after the peak load, and consists of the growth of the curve through the loss of stiffness due to plastic damage. The relationship that governs this stretch is tensile stress and inelastic strain, which are input data from the model.

Equations (3.7) and (3.8) calculate inelastic deformation. Finally, Eq. (3.9) calculates plastic deformation.

$$\varepsilon_t^{\sim ck} = \varepsilon_t - \varepsilon_{t0}^{el} \tag{3.7}$$

$$\varepsilon_{t0}^{el} = \frac{\sigma_t}{E_0} \tag{3.8}$$

$$\varepsilon_t^{\sim pl} = \varepsilon_t^{\sim ck} - \frac{d_t}{(1 - d_t)} \frac{\sigma_t}{E_0} \tag{3.9}$$

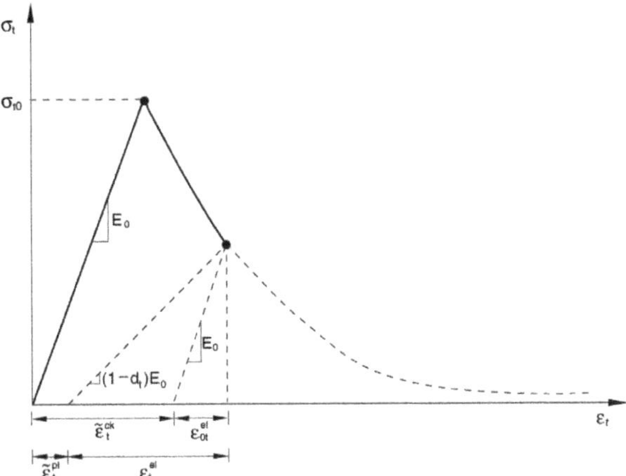

Fig. 3.2 Definition of traction inelastic and plastic deformation (Hibbit et al. 2012)

3.2.3 Nonlinear Behaviour in Compression

The nonlinear behaviour in traction is represented in Fig. 3.3. The nonlinear stretch begins after the load of the elastic limit (σ_{c0} in Fig. 3.3), and grows to the peak load, where after the curve growth through the loss of stiffness due to plastic damage. The relationship that governs this stretch is tensile stress and inelastic strain, which are input data from the model.

Equations (3.10) and (3.11) calculate inelastic deformation. Equation (3.12) calculates plastic deformation.

$$\varepsilon_c^{\sim in} = \varepsilon_c - \varepsilon_{c0}^{el} \tag{3.10}$$

$$\varepsilon_{c0}^{el} = \frac{\sigma_c}{E_0} \tag{3.11}$$

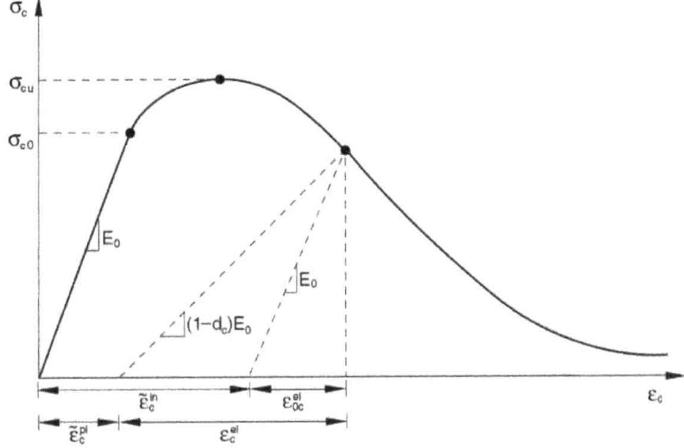

Fig. 3.3 Definition of inelastic and plastic compression deformation (Hibbit et al. 2012)

$$\varepsilon_c^{\sim pl} = \varepsilon_c^{\sim in} - \frac{d_c}{(1 - d_c)} \frac{\sigma_c}{E_0} \tag{3.12}$$

3.3 Flow Law and Flow Criterion

The law of flow is the hyperbolic function of Drucker-Prager, which is represented by Eq. (3.13), and presented in Fig. 3.4. Equation (3.14) presents the effective hydrostatic stress, $\underline{p}$, while Eq. (3.15) the effective Mises equivalent equation, $\underline{q}$, and Eq. (3.16) the effective tensor voltage, $\underline{S}$.

$$G = \sqrt{(\epsilon \cdot \sigma_{t0} \cdot \tan \psi) + \underline{q}^2} - \underline{p} \cdot \tan \psi \tag{3.13}$$

$$\underline{p} = -\frac{1}{3} traco\ \underline{\sigma} \tag{3.14}$$

$$\underline{q} = \sqrt{\frac{3}{2}(\underline{S} : \underline{S})} \tag{3.15}$$

$$\underline{S} = \underline{\sigma} + p \cdot I \tag{3.16}$$

In the above equations, ϵ is eccentricity, σ_{t0} is peak stress in traction, the angle of dilation, σ is the effective stress and I is the stress invariant.

Eccentricity is how much the Drucker-Prager function approaches asytomytically from its straight projection (dashed lines in Fig. 3.4). According to Hibbit et al. (2012), if eccentricity is zero the function tends to be a line, and advise the use of the value 0.1 as standard. A relevant contribution is by Jankowiak and Lodygowski (2005) who indicate a way to calculate the parameter, using the ratio between tensile strength and compressive strength. By uniting the contributions, it is possible to infer that if the eccentricity value was 0.1, the tensile strength value has to be close to 10% of the

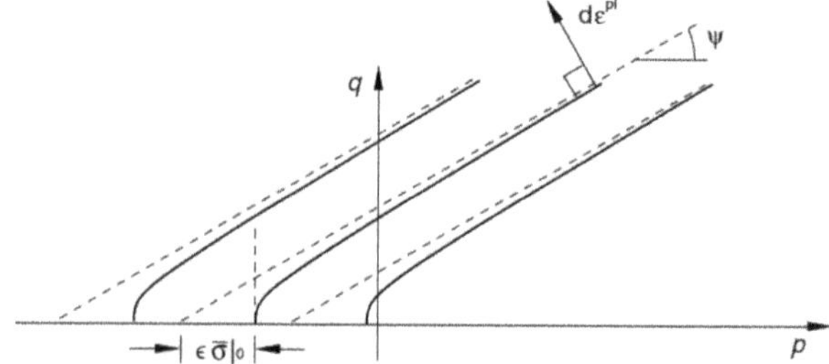

Fig. 3.4 Hyperbolic approach of Drucker-Prager plasticization function (Hibbit et al. 2012)

compressive strength value. According to Soares (2016), the increase in eccentricity would imply a greater curvature of the Drucker-Prager function by increasing the dilation angle as the confinement stress decreases, and on the contrary, the reduction can cause convergence errors if the concrete is subjected to low confinement pressure. The author goes on to say that the value of 0.1 implies that for variable levels of confinement stress the dilation angle remains almost immutable.

As Fig. 3.4 presents, the dilation angle governs the slope of the flow surface along the hydrostatic axis in the southern plane. For Kmiecik and Kaminski (2011), one way of physically understanding the angle of dilation is as the internal friction angle of the concrete, thus being its value between 36° and 40°. Soares (2016) contributes by saying that if the dilation angle is low, the concrete will present a more fragile behaviour, already for high values the behaviour will be dubious. Based on the knowledge that concrete has a near-fragile behaviour, also based on the research of Vermeer and Borst (1984), Alfarah et al. (2017) defends the value of 13°. If the dilation angle is zero the material does not dilate, and the maximum value accepted by ABAQUS is 56.3°. For Doroszko and Seweryn (2015), high dilation angle values will bring positive volumetric deformation responses in the compression zone, which will generate an unrealistic increase in load capacity; the authors recommend the use of 5° as an angle of dilation. The researchers who bring their contributions on the CDP agree that one of the challenges of using the model is the calibration of the dilation angle, since in the literature there is a high variance among the researchers, being in charge of each research to seek their strategies.

The flow criterion is guided by the flow function of Lubliner et al. (1989), with modifications that were coupled by Lee and Fenves (1998). The flow surface has the evolution governed by plastic tensile and compression deformations. Equation (3.17) presents the function of Lubliner et al. (1989) with the modifications of Lee and Fenves (1998).

Equations (3.18)–(3.20) calculate variables that are dimensional constants determined using experimental data. The variable $\propto$ is the ratio between biaxial compression resistance and elastic compression limit tension; its value is between 0 and 0.5, dependent on σ_{b0}/σ_{c0} has as standard value 1.16, which means that the biaxial compressive strength is 16% higher than the uniaxial. The variable β depends beyond $\propto$, on the effective cohesive tension in compression and traction. Alfarah et al. (2017) points out that these stresses are $\sigma_i/(1-d_i)$, being i suffix to indicate traction (t) or compression (c).

$$F = \frac{1}{1-\alpha}\left(\underline{q} - 3\alpha\underline{p} + \beta\left(\varepsilon^{\sim pl}\right)<\underline{\widehat{\sigma_{máx}}}> - \gamma<\underline{\widehat{\sigma_{máx}}}>\right) - \underline{\sigma_c}\left(\varepsilon_c^{\sim pl}\right) = 0 \quad (3.17)$$

with

$$\alpha = \frac{\sigma_{b0}/\sigma_{c0} - 1}{2(\sigma_{b0}/\sigma_{c0}) - 1} \quad (3.18)$$

$$\beta = \frac{\bar{\sigma}_c\left(\tilde{\varepsilon}_c^{pl}\right)}{\bar{\sigma}_t\left(\tilde{\varepsilon}_t^{pl}\right)}(1-\alpha) - (1+\alpha) \tag{3.19}$$

$$\gamma = \frac{3(1-K_c)}{2K_c - 1} \tag{3.20}$$

According to Soares (2016), the variable γ defines the format of loading on the flow surface. This variable depends on the K_c which can be understood as the ratio in the derailing plane of the distance of the hydrostatic axis and the tensile and compression meridians. Alfarah et al. (2017) points out that K_c can range from 0.5 (Rankine surface) and 1 (Von Mises surface); the authors follow the surface of Mohr–Coulomb ($K_c = 0.7$). Hibbit et al. (2012) suggests the value of 2/3 for K_c. Figure 3.5 shows the flow surfaces in the diverter plane for the aforementioned K_c.

Figure 3.6 is interesting because it shows the behaviour of the run-off surface in the concrete phases.

The CDP takes into account the linear and nonlinear behaviour of concrete, where in softening stiffness is degraded. It is at this stage of the simulation that errors in modelling may occur due to localized deformation. This localized deformation can be understood as a discontinuity generated by the criterion of flow having been reached at certain points, causing there to be places that are in the linear behaviour, while others are in the nonlinear behaviour.

One possible solution to overcome this problem is the use of the visco-plastic regularization method, as suggested by Duvaut-Lions apud Hibbit et al. (2012). The use of this method happens when the viscosity parameter is inserted; thus, the plastic deformations and damage are smoothed by the viscosity coefficient. Soares (2016) states that the use of small viscosity values helps in the convergence rate in the softening regime.

It is worth noting that if the viscosity is zero the visco-plastic regularization will not be used. According to Doroszko and Seweryn (2015), the viscosity value refers to relaxation time in seconds, and that the use of high viscosity values causes the damage

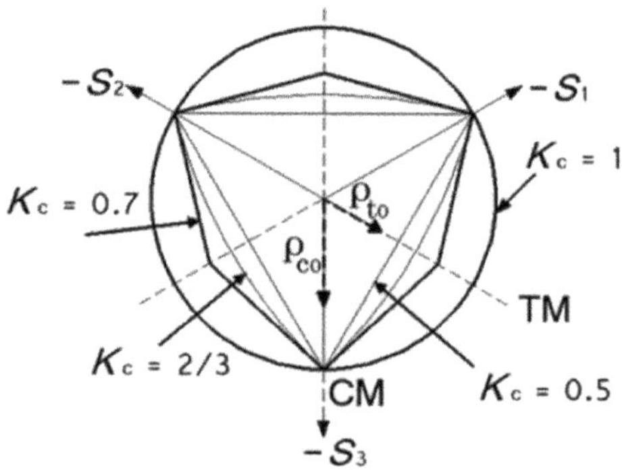

Fig. 3.5 Flow surface in the diverter plane for Kc values (Hibbit et al. 2012)

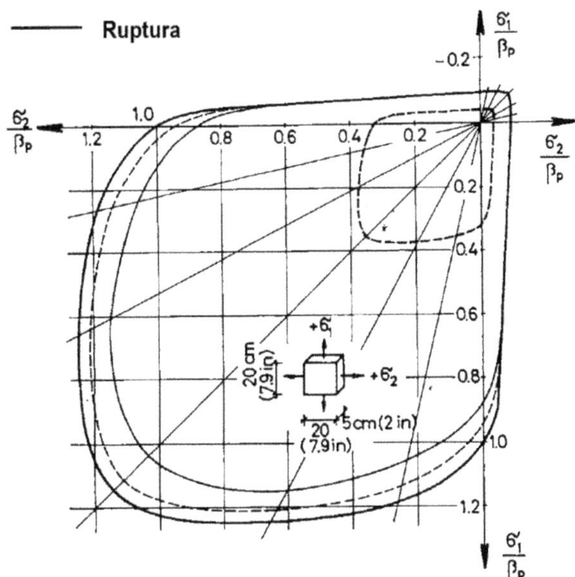

Fig. 3.6 Flow surface in the stress plane (Kupfer et al. 1969)

zone to spread through several finite elements. The authors continue to explain that this behaviour limits the propagation of fissures and makes the pattern of the same.

One point that takes a lot of attention from researchers is the parameters of damage in compression and traction. These parameters work with the input of stress, inelastic strain and damage data; ABAQUS transforms inelastic deformation into plastic deformation and thus is done all the criterion of plastic damage, where the concrete loses stiffness as the damage increases.

This point proved particularly difficult to obtain in the literature, which encouraged the creation of a routine that was able to generate the data necessary for the use of the parameters of damage in compression and traction. The next section will deal with the routine elaborated in the MATLAB program.

3.4 Routine for Parameters Needed for the Model CDP

For the generation of plasticity parameters to be inserted in the CDP model, a routine at MATLAB was created, based on the work of Alfarah et al. (2017) and its main features are as follows:

- Modifies the formulation of Lubliner et al. (1989) and Lee and Fenves (1998) to obtain the damage variables in relation to the corresponding deformations, through the integration of fracture and crushing energies of the concrete.

- Excludes the need for experimental calibration. This feature is achieved through an interactivity, where an average value is used for the ratio between plastic compression deformation and elastic compression deformation.
- The routine is simple to run, since it only requires three input parameters.

The contribution of this research was to obtain the mechanical properties from the ACI-318 (2014), since the basis of Alfarah's work is the FIB (2010). This allowed the availability of two hypotheses to obtain the mechanical properties of the concrete, an aspect that increases the probability of adequacy of the numerical model.

3.4.1 Routine Description

The initial definition of damage is modified from the Lee and Fenves equation (1998), and is presented in Eqs. (3.21) and (3.22). As mentioned above, ε_c^{ch} and ε_t^{ck} are respectively, elastic deformation of compression and tensile (or crushing and fissure) that are replaced by their plastic components originally present in the equations.

$$dc = \frac{1}{gc}\int_0^{\varepsilon_c^{ch}} \sigma_c \; d\varepsilon_c^{ch} \tag{3.21}$$

$$dt = \frac{1}{gt}\int_0^{\varepsilon_t^{ck}} \sigma_t \; d\varepsilon_t^{ck} \tag{3.22}$$

The *gc* and *gt* variables are the energies dissipated by the damage per unit volume along the loss of stiffness. They are the equivalent, in ABAQUS, to the energy dissipated by volume by the damage. The formulas for these energies are presented in Eqs. (3.23) and (3.24).

$$gc = \int_0^{\infty} \sigma_c d\varepsilon_c^{ch} \tag{3.23}$$

$$gt = \int_0^{\infty} \sigma_t d\varepsilon_t^{ck} \tag{3.24}$$

Equations (3.23) and (3.24) can be understood graphically as the integral of the *Stress versus Inelastic Deformation curve*. You can relate the energies per volume unit with the energies per unit area using the equivalent length of the finite element mesh (l_{eq}). Equations (3.25) and (3.26) present this relationship.

The variables G_{ch} and G_f can be defined as crushing energy and fracture energy. The tension equation presented by Lubliner et al. (1989) was modified by Lee and Fenves (1998) and generate two separate equations for compression and traction tension.

$$gc = \frac{G_{ch}}{l_{eq}} \tag{3.25}$$

$$gt = \frac{G_f}{l_{eq}} \tag{3.26}$$

The proposal of Alfarah et al. (2017) was to modify the equation of Lee and Fenves (1998), modifying the plastic deformations into inelastic. Equations (3.27) and (3.28) present the result of these modifications.

$$\sigma_c = f_{c0}\left[(1 + a_c)\exp\left(-b_c \cdot \varepsilon_c^{ch}\right) - a_c \cdot \exp\left(-2b_c \cdot \varepsilon_c^{ch}\right)\right] \tag{3.27}$$

$$\sigma_t = f_{t0}\left[(1 + a_t)\exp\left(-b_t \cdot \varepsilon_t^{ck}\right) - a_t \cdot \exp\left(-2b_t \cdot \varepsilon_t^{ck}\right)\right] \tag{3.28}$$

The variables f_{c0} and f_{t0} are the flow limit stresses. For the stress level corresponding to these stresses, the inelastic deformations must be null because the material is still in the linear regime.

There are four dimensional variables in Eqs. (3.27) and (3.28), i.e.: a_c, a_t, b_c and b_t. By replacing Eqs. (3.27) and (3.28), respectively, into Eqs. (3.23) and (3.24) it is possible to obtain a new expression for g_c and g_t that are written according to Eqs. (3.29) and (3.30) presented below.

$$gc = \frac{f_{c0}}{b_c}\left(1 + \frac{a_c}{2}\right) \tag{3.29}$$

$$gt = \frac{f_{t0}}{b_t}\left(1 + \frac{a_t}{2}\right) \tag{3.30}$$

Replacing Eqs. (3.25) and (3.26), respectively, into Eqs. (3.29) and (3.30), it is possible to obtain the values of b_c and b_t. Equations (3.31) and (3.32) show the formulation for obtaining the variables mentioned above.

$$b_c = \frac{f_{c0} \cdot l_{eq}}{G_{ch}}\left(1 + \frac{a_c}{2}\right) \tag{3.31}$$

$$b_t = \frac{f_{t0} \cdot l_{eq}}{G_f}\left(1 + \frac{a_t}{2}\right) \tag{3.32}$$

To obtain the equation of damage in compression and traction, Eqs. (3.27) and (3.29) are replaced into Eq. (3.21) and Eqs. (3.28) and (3.30) into Eq. (3.22). The results of the substitutions will be shown in Eqs. (3.33) and (3.34), which are the basic equations for obtaining the compression and traction damage used in the routine.

$$d_c = 1 - \frac{1}{2 + a_c}\left[2(1 + a_c)\exp\left(-b_c \cdot \varepsilon_c^{ch}\right) - a_c \cdot \exp\left(-2b_c \cdot \varepsilon_c^{ch}\right)\right] \tag{3.33}$$

$$d_t = 1 - \frac{1}{2 + a_t}\left[2(1 + a_t)\exp\left(-b_t \cdot \varepsilon_t^{ck}\right) - a_t \cdot \exp\left(-2b_t \cdot \varepsilon_t^{ck}\right)\right] \tag{3.34}$$

Equalling zero the derivatives of σ_c and σ_t in relation to, respectively, ε_c^{ch} and ε_t^{ck}, in Eqs. (3.27) and (3.28), it is possible to calculate peak stresses in compression and traction. Equations (3.35) and (3.36) present the calculation of these maximum stresses.

$$f_{cm} = \frac{f_{c0}(1 + a_c)^2}{4a_c} \tag{3.35}$$

$$f_{tm} = \frac{f_{t0}(1 + a_t)^2}{4a_t} \tag{3.36}$$

From Eqs. (3.35) and (3.36) it is possible to obtain the variables a_c and a_t, highlighting them in the equations in question. Equations (3.37) and (3.38) present the formulas for obtaining, respectively, a_c and a_t.

$$a_c = 2(f_{cm}/f_{c0}) - 1 + 2\sqrt{\left(\frac{f_{cm}}{f_{c0}}\right)^2 - \left(\frac{f_{cm}}{f_{c0}}\right)} \tag{3.37}$$

$$a_t = 2(f_{tm}/f_{t0}) - 1 + 2\sqrt{\left(\frac{f_{tm}}{f_{t0}}\right)^2 - \left(\frac{f_{tm}}{f_{t0}}\right)} \tag{3.38}$$

The bases for the calculations of the damage parameters are contained in Eqs. (3.31)–(3.38). These equations are dependent on the variables f_{cm}, f_{tm}, f_{c0}, f_{t0}, G_{ch}, G_f, ε_c^{ch}, ε_c^{ch} and l_{eq}. At this point it is worth the memory of Eqs. (3.7) and (3.10) that define ε_c^{ch} and ε_c^{ch}. It is also necessary to summarize the concept of deformations so that their differences in the CDP are clear and the strategies used are understood.

CDP unites the concepts of the plasticity and damage model. Concrete is a material of more fragile behaviour, but in fields where there is an inversion of stresses the microcracks can close and there may be redistribution, causing the damage model to represent well the behaviour of the concrete. Unlike concrete, in steel it is not possible to have redistribution and there are rare cracks, having a markedly dubious behaviour and the plasticity model is appropriate. Thus, CDP is a powerful tool to represent the behaviour of concrete, especially reinforced concrete.

Figure 3.7 presents the graphical representation of the models of damage, plasticity and the coupling of both—o CDP.

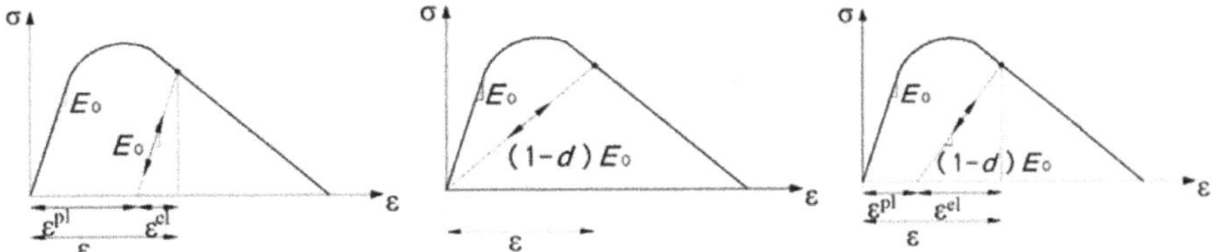

Fig. 3.7 Representation of CDP and its root models in compression (Alfarah et al. 2017). **a** Plasticity model, **b** damage model and **c** plastic damage model

Analysing Fig. 3.7, it is possible to understand the compositions of deformations in each root model and thus understand the intrinsic concepts in the CDP. In the plasticity model the total deformation (ε) can be decomposed into plastic deformation and elastic deformation, respectively, ε^{pl} and ε^{el}. In the damage model, the deformations come from damage in material stiffness and are irreversible. When the two concepts are united, the decomposition of the deformations goes through changes that can bring some confusion. Figure 3.8 helps in understanding the concepts discussed.

By joining both models, the plastic deformation of the plasticity model is decomposed into inelastic deformation (in compression can be defined as crushing and traction can be defined as fissure) and plastic deformation. It is important to understand that plastic deformation is contained in inelastic deformation, being called b the variable that relates the two deformations. Equation (3.39) presents this generalized relationship for compression and traction.

$$b = \frac{\varepsilon^{pl}}{\varepsilon^{in}} \tag{3.39}$$

From Fig. 3.8, the concept of undamaged and damaged elastic deformation will be defined (in this work differentiated as, respectively, ε_0^{el} and ε^{el}). The first stage of concrete is linear and elastic (stresses up to $\underline{\sigma}_{c0}$ in the figure). This means that the

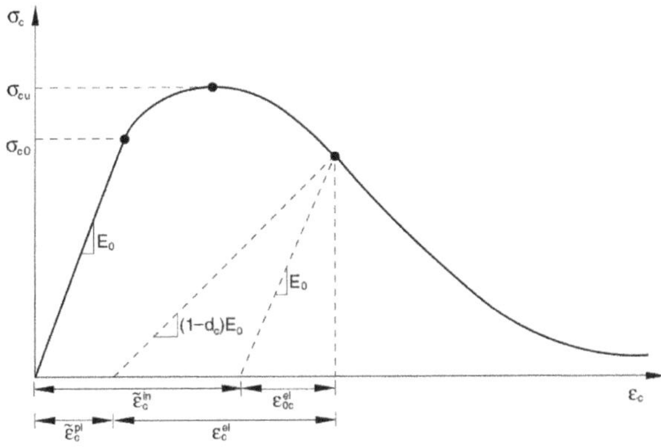

Fig. 3.8 Definition of inelastic and plastic compression deformation (Hibbit et al. 2012)

deformations are completely reversible in the event of a discharging process. These deformations are undamaged elastic deformations (ε_{0c}^{el} in Fig. 3.8).

There is a distinction between compression and traction in the second stage, because in traction this stage is already softening and there is damage to stiffness. In compression, stage two is the ascending nonlinear stage that goes up to the peak voltage (σ_{cu} in Fig. 3.8). At this stage there are inelastic compression deformations (or crushing deformations), however the damage in compression is null or very small. The softening stage (stage two in traction and stage three in compression) occurs after the peak voltage is reached. At this stage there is damage to stiffness, however there is the emergence of a deformation component that is reversible in discharging, but which is not fully recoverable because it suffers from the damage in stiffness. This deformation is elastic (since it can be recoverable) but has been damaged by damage in stiffness. This is the concept of damaged elastic deformation (ε_{c}^{el} in Fig. 3.8).

Based on the above explanation, the concept of inelastic deformation is one that includes a non-recoverable deformation (plastic deformation) and a recoverable but damaged deformation (damaged elastic deformation). Thus, the fact that ABAQUS requires stresses, inelastic deformations and damage is based on the possible stages of decomposition of the deformations. From the total deformation is removed the undamaged elastic portion, what remains is the inelastic deformation. From this, the damaged elastic portion is removed by remaining plastic deformation. Equation (3.40) summarizes these decompositions.

$$\varepsilon - \varepsilon_0^{el} = \varepsilon^{in} \rightarrow \varepsilon^{in} - \varepsilon^{el} = \varepsilon^{pl} \tag{3.40}$$

The definitions of total deformations are fundamental for the calculation of inelastic deformations and damage. The *stress versus deformation* behaviour of concrete can be defined by various standards and codes, for example, in the work presented by Alfarah et al. (2017) is based on the FIB (2010) for the upward stretch of the curve in compression and the softening is based on the work of Krätzig and Pölling (2004). In the traction the authors follow the concepts contained in the works of Vonk (1993) and Van Mier (1986) and the descending stretch of the curve is based on the concept of fissure opening and a kinematic relationship.

The three excerpts in compression are presented in Fig. 3.9 as the respective Eqs. (3.41)–(3.43).

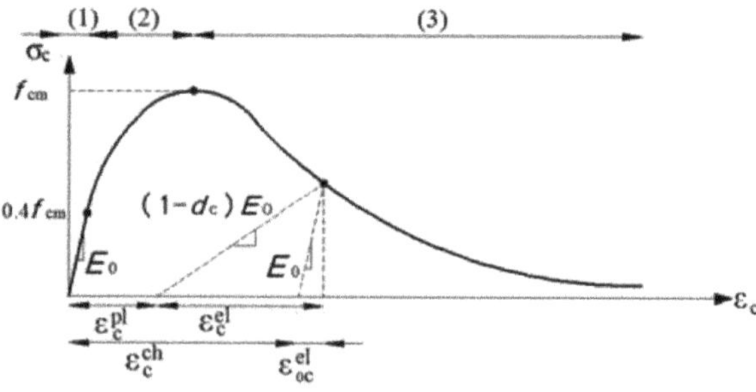

Fig. 3.9 Concrete behavior in compression (Alfarah et al. 2017)

$$\sigma_{c1} = \varepsilon_c \cdot E_0 \tag{3.41}$$

$$\sigma_{c2} = \frac{E_{ci} \cdot \frac{\varepsilon_c}{f_{cm}} - \left(\frac{\varepsilon_c}{\varepsilon_{cm}}\right)^2}{1 + \left(E_{ci} \cdot \frac{\varepsilon_{cm}}{f_{cm}} - 2\right) \cdot \frac{\varepsilon_c}{\varepsilon_{cm}}} \cdot f_{cm} \tag{3.42}$$

$$\sigma_{c3} = \left(\frac{2 + \gamma_c \cdot f_{cm} \cdot \varepsilon_{cm}}{2 \cdot f_{cm}} - \gamma_c \cdot \varepsilon_c + \frac{\varepsilon_c{}^2 \cdot \gamma_c}{2 \cdot \varepsilon_{cm}}\right)^{-1} \tag{3.43}$$

where E_0 is the secant modulus of elasticity, E_{ci} is the tangent modulus of elasticity, f_{cm} is the peak voltage in compression, γ_c is a constant that, according to Krätzig and Pölling (2004), controls the area of the softening stretch (this area is finite and governed by the G_{ch} and l_{eq}) and ε_{cm} is the total deformation related to peak voltage, its approximate value is 0.0022.

Equations (3.44)–(3.48) present the formulas for obtaining the above mentioned variables.

$$E_0 = \left(0.8 + \frac{0.2 \cdot f_{cm}}{88}\right) \cdot E_{ci} \tag{3.44}$$

$$E_{ci} = 10000 \cdot f_{cm}^{\frac{1}{3}} \tag{3.45}$$

$$f_{cm} = f_{ck} + 8 \tag{3.46}$$

$$\gamma_c = \frac{\pi^2 \cdot f_{cm} \cdot \varepsilon_{cm}}{2\left[\frac{G_{ch}}{l_{eq}} - 0.5 \cdot f_{cm} \cdot \left(\varepsilon_{cm} \cdot (1-b) + b \cdot \frac{f_{cm}}{E_0}\right)\right]^2} \tag{3.47}$$

$$\varepsilon_{cm} = \frac{2 \cdot f_{cm}}{E_0} \tag{3.48}$$

The two traction segments are presented in Fig. 3.10 as the respective Eqs. (3.49) and (3.50).

$$\sigma_{t1} = \varepsilon_t \cdot E_0 \tag{3.49}$$

$$\sigma_{t2}(w) = \left[\left[1 + \left(c1 \cdot \frac{w}{w_c}\right)^3\right] \cdot e^{\left(-c2 \cdot \frac{w}{w_c}\right)} - \frac{w}{w_c} \cdot (1 + c1^3) \cdot e^{(-c2)}\right] \cdot f_{tm} \tag{3.50}$$

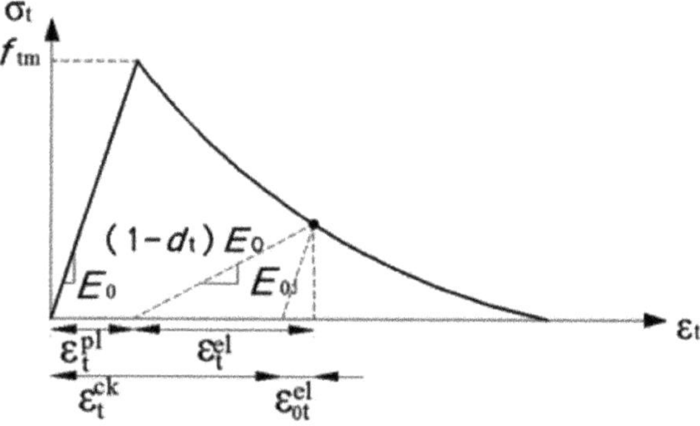

Fig. 3.10 Concrete behavior in traction (Alfarah et al. 2017)

where $c1 = 3.0$ and $c2 = 6.93$, w it's fissure opening and w_c is the maximum crack opening.

When the concrete reaches w_c as a crack opening, there is a complete rupture of the material. In this way, $\sigma_{t2}(0) = f_{tm}$ and $\sigma_{t2}(w_c) = 0$. Equation (3.51) displays the w_c value as a function of fracture energy (G_f).

$$w_c = 5.14 \cdot \frac{G_f}{f_{tm}} \tag{3.51}$$

The total tensile deformation (ε_t) is calculated with a ratio to the crack opening (w) and the limit deformation of the first stretch in the traction (ε_{tm}), which is also the deformation of the maximum tensile stress (f_{tm}). The following equations present the formulas of the above mentioned variables.

$$\varepsilon_t = \varepsilon_{tm} + \frac{w}{l_{eq}} \tag{3.52}$$

$$\varepsilon_{tm} = \frac{f_{tm}}{E_0} \tag{3.53}$$

$$f_{tm} = 0.3016 \cdot fck^{\frac{2}{3}} \tag{3.54}$$

Finally, it is necessary to define the fracture and crushing energies (respectively, G_f and G_{ch}). Equations (3.55) and (3.56) allow the achievement of these energies and are those that appear, respectively, in FIB (2010) and Oller's work (1988).

$$G_f = 0.073 \cdot f_{cm}^{0.18} \tag{3.55}$$

$$G_{ch} = \left(\frac{f_{cm}}{f_{tm}}\right)^2 \cdot G_f \tag{3.56}$$

3.4.2 Routine Implementation—Generation of Inelastic Deformations, Damage Parameters and Respective Stresses

The routine was developed in MATLAB and its script can be found in Appendix A. Following are the steps to be followed to obtain the inelastic properties and damage in the CDP model.

- Enter the values of the characteristic compressive strength (f_{ck} in MPa), the characteristic length of the finite element mesh (l_{eq} in mm) and the relationship between plastic and inelastic deformation in compression (b, Eq. 3.39). The initial value of b is estimated and will be found interactively its average value.
- Calculate peak stresses in compression and traction (f_{cm} and f_{tm}, Eqs. 3.46 and 3.54)
- Calculate the modulus of tangent and secant elasticity (E_{ci} and E_0, Eqs. 3.45 and 3.44)
- Calculate peak deformations in compression and traction (ε_{cme} and ε_{tm}, Eqs. 3.48 and 3.53)
- Calculate fracture and crush energy (G_f and G_{ch}, Eqs. 3.55 and 3.56)
- Calculate the maximum crack opening (w_c, Eq. 3.51)
- Assemble the *stress versus strain curves* in compression, Eqs. 3.41, 3.42 and 3.43; as mentioned above, the curve is limited, the value of the last total deformation (ε_{cu}) must be estimated so that the area of the curve multiplied by the *leq* is equal to G_{ch}. The strategy in compression is to generate values for the total deformation (ε_c) and thus calculate their respective stresses
- Assemble the *stress versus strain curves* in traction, Eqs. (3.49) and (3.50). The strategy in traction, in the first stretch the strategy is the same as compression, however in the second stretch crack opening values are generated (w in mm) remembering that it has a spread of 0 to w_c, and then the total deformations (ε_t) of the second stretch are calculated using Eq. (3.52)
- Calculate inelastic deformations in compression and traction (Eq. 3.40)
- Calculate damage parameters a_c, a_t, b_c and b_t by Eqs. (3.37), (3.38), (3.31) and (3.32)
- Calculate compression and traction damage variables (Eqs. 3.33 and 3.34)
- Calculating plastic deformations in compression and traction (Eqs. 3.12 and 3.9)
- Calculate the mean value of b in compression (Eq. 3.39), being different from the estimated value, repeat the entire process with b until convergence.

3.5 Routine for Reducing the Mechanical Properties of Concrete by Internal Expansion Reactions (IER)

The inelastic properties and damage of the CDP Model depend on the mechanical properties of the concrete that were affected by internal expansion reactions. With

Table 3.1 Reduction rates in mechanical properties according to the level of expansion

Distress mechanism	Reference expansion level (%)	Damage results		
		Stiffness loss (%)	Compressive strength loss (%)	Tensile strength loss (%)
ASR 01 (Zhang et al. 2019)	0.04 ± 1	30	60	10
	0.11 ± 1	45	65	15
	0.20 ± 1	60	70	25
	0.30 ± 1	65	80	35
ASR 02 (Blanco et al. 2018)	0.04 ± 1	37	60	15
	0.11 ± 1	50	65	20
	0.20 ± 1	60	80	25
	0.30 ± 1	67	80	25
ASR 03 (Zhang et al. 2019)	0.05	18	25	2
	0.10	29	40	12
	0.15	38	48	15

this fact in mind, a way to implement the reductions in the mechanical properties of concrete, reported by Sanchez et al. (2017, 2018) was developed in the CDP routine. For this purpose, an additional routine was developed for this purpose, called CDP_RIE.

The strategy chosen was a qualitative selection of Sanchez's results, selecting the values presented in Sanchez et al. (2017) work (RIE_01) and the results of a concrete made with a large aggregate from Recife-PE (RIE_03). From Sanchez et al. (2018), the values referring to RAS (RIE_02) and the values referring to DEF (RIE_04) were selected. Table 3.1 presents a summary of the values adopted.

References

Alfarah B, López-Almansa F, Oller S (2017) New methodology for calculating damage variables evolution in plastic damage model for RC structures. Eng Struct 132:70–86

Blanco A, Cavalaro SHP, Segura I, Segura-Castillo L, Aguado A (2018) Expansions with different origins in a concrete dam with bridge over spillway. Constr Build Mater 163:861–874

Doroszko M, Seweryn A (2015) Modeling of the tension and compression behavior of sintered 316L using micro computed tomography. Acta Mechanica et Automatica 9:70–74

Hibbit KS (2012) ABAQUS/Standard User's manual. Pawtucket, 12 Editon

Jankowiak T, Lodygowski T (2005) Identification of parameters of concrete damage plasticity constitutive model. Found Civ Environ Eng 6(1):53–69

Kmiecik P, Kamiński M (2011) Modelling of reinforced concrete structures and composite structures with concrete strength degradation taken into consideration. Arch Civ Mech Eng 11(3):623–636

Krätzig WB, Pölling R (2004) An elasto-plastic damage model for reinforced concrete with minimum number of material parameters. Comput struct 82(15–16):1201–1215

Kupfer H, Hilsdorf HK, Rusch H (1969) Behavior of concrete under biaxial stresses. J Proc 656–666

Lubliner J, Oliver J, Oller S, Oñate E (1989) A plastic-damage model for concrete. Int J Solids Struct 25(3):299–326

Lee J, Fenves GL (1988) Plastic-damage model for cyclic loading of concrete structures. J Eng Mech 124(8):892–900

Sanchez LFM, Fournier B, Jolin M, Mitchell D, Bastien J (2017) Overall assessment of alkali-aggregate reaction (AAR) in concretes presenting different strengths and incorporating a wide range of reactive aggregate types and natures. Cem Concr Res 93:17–31

Soares LMS (2016) Nonlinear numerical analysis of pillar connections fungiform slab, MSc. Thesis, Universidade Nova de Lisboa, Faculdade de Ciências e Tecnologia

Sanchez LFM, Drimalas T, Fournier B, Mitchell D, Bastien J (2018) Comprehensive damage assessment in concrete affected by different internal swelling reaction (ISR) mechanisms. Cem Concr Res 107:284–303

van Mier JGM (1986) Multiaxial strain-softening of concrete. Mater Struct 19(3):190–200

Vermeer PA, De Borst R (1984) Non-associated plasticity for soils concrete and rock. Heron 29(3):1–64

Vonk RA (1993) A micromechanical investigation of softening of concrete loaded in compression. Heron 38:3–94

Zhang H, Li L, Wang W (2019) Effects of temperature rising inhibitor on nucleation and growth process of ettringite. J Solid State Chem 274:222–228

Chapter 4
Bottle-Shaped Isolated Struts Concrete Deteriorated

Sankovich's research analysed the behaviour of bottle-shaped connecting rods using simple and reinforced concrete panels with various transverse reinforcement configurations, in which the compressive stresses resulting from the load applied through rigid plates installed at the top and base of the elements were able to disperse creating transverse tensile stresses within the samples.

Sankovich's study was fundamentally developed to evaluate the recommendations regarding compression stress limits and frame demands in bottle-shaped connecting rods recommended by ACI 318-02 (2002) and AASHTO LRFD (1988) with the aim of unifying the treatment of connecting rods in the form of bottles in these two important standards.

In the next subsections will be described the experiments carried out focusing on those that are of interest to the study carried out, i.e., simple concrete panels.

4.1 Materials and Methods

In order to understand the behaviour of isolated connecting rods of bottle-shaped concrete deteriorated by internal expansion reactions, numerical analyses were performed in a linear and nonlinear regime on the same panels tested by Sankovich (2003) to, in an initial approach, promote the validation of the numerical representation model to be used—the Concrete Damaged Plasticity Model of ABAQUS.

Partially loaded unreinforced concrete panels with dimensions of 914.4 × 914.4 × 152.5 mm^3 were modelled, as well as the 304.8 × 152.5 × 50.8 mm^3 steel plates at its top and base as it is shown in Fig. 4.1. Details of the two panels modelled are summarized in Table 4.1.

To obtain the properties necessary for the characterization of tensile and compression damage of the concrete, routines were developed in the MATLAB (2012) that

A. C. Azevedo et al., *Concrete Structures Deteriorated by Delayed Ettringite Formation and Alkali-Silica Reactions*, Building Pathology and Rehabilitation 24,
https://doi.org/10.1007/978-3-031-12267-5_4

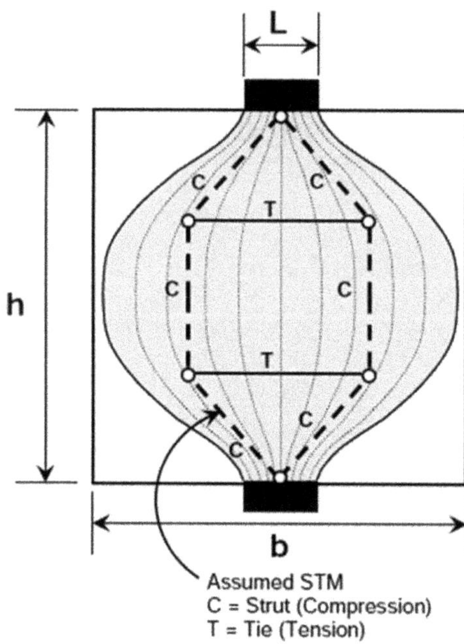

Fig. 4.1 Bottle-shaped struts and associated strut and tie model

Table 4.1 Tested result of specimens S1-2 and S3-1

Specimen	Specimen dimensions (mm)	Load plate dimensions (mm)	f′c (MPa)	First cracking load (kN)	Failure load (kN)	Principal tensile strain (ε_1)
S1-2	914.4 × 914.4 × 101.6	304.8 × 50.8 × 101.6	26.41	524.0	706.8	0.0022
S3-1	914.4 × 914.4 × 152.4	304.8 × 50.8 × 152.4	28.96	608.5	873.2	0.0014

allow the obtaining of the necessary parameters with few input data, i.e., the characteristic strength to the concrete compression, the equivalent length of the finite element mesh and the relationship between plastic deformation and inelastic deformation in compression. The routines allow the choice of two methods for the calculation of damage in concrete and two possibilities for calculating mechanical properties—ACI 318–14 (2014) and FIB (2010).

Consideration of the effects of internal expansion reactions on concrete was introduced from Sanchez's work (2014, 2015, 2016, 2017 and 2018). Sanchez's research indicated that the action of internal expansion reactions has as an important consequence the reduction of the properties of resistance and deformation of concrete. Since these reductions directly affect the inelastic deformations and the parameters of concrete damage, both in traction and compression regime (the base of the CDP), it was decided to elaborate a specific routine in the MATLAB in which the input parameters are the values of the observed expansions. The consequent change in the resistance and deformation properties of the concrete is automatically calculated from the routine.

With the CDP model duly calibrated and validated and with the data of change of the mechanical properties of the concrete from the level of expansion was conducted a representative scope of numerical analyses with various scenarios, namely: panel fully affected by internal expansion reactions, panel with effects of expansion located only in the connecting rod in bottle form, panel with change of all mechanical properties simultaneously and panels with isolated changes of mechanical properties. The geometry of the connecting rod used was defined from numerical analyses in linear regime of the panel investigated. The specific details, as well as the details of the hypotheses adopted in the routine and details of the modelling are discussed in subsequent specific sub-chapters.

4.2 Experimental Program

In all, 28 simple and reinforced concrete panels were tested. The tests performed are analogous to diametric compression tests in cylindrical concrete specimens, in which a monotonic load is increasingly applied to the sample whose geometry and orientation of the loading allowed the dispersion of compression and tensile stresses inside the sample.

The load was applied with a universal test machine through steel plates smaller than the panels to create a partial loading situation, identical to partially loaded blocks. Figure 4.2 shows the configuration of the tests performed.

To ensure that the force from the cell was applied to the concrete surface uniformly, a material of specific characteristics—hydrostone, between the steel plate and the concrete, was used and a spherical "head" was installed at the interface of the loading lava. Each tested panel has been rectified, at the top and base, to ensure its perfect alignment and levelling.

The deformations were measured using strain gauge that were placed or analysed in the reinforcements on the concrete surface. Data were collected through the user's acquisition systems available at the Ferguson Structural Engineering Laboratory at the University of Texas at Austin. Displacement records were not made, since Sanders (1990) and Wollmann (1992) in their experiments concluded that panels like these ruptured in a non-dubious manner, providing few or no useful displacement versus load data.

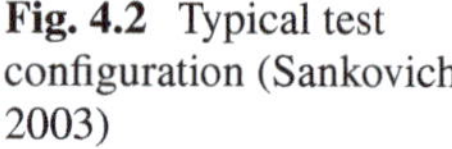

Fig. 4.2 Typical test configuration (Sankovich 2003)

4.2.1 *Panel S1-2*

The dimensions of this panel were as follows: 914.4 × 914.4 × 101.6 mm, with steel plates at the top and base dimensions of 304.8 × 50.8 × 101.6 mm. The instrumentation of this panel was performed as shown in Fig. 4.3 In it can be observed the rosettes with strain gauges arranged at angles of 0°, 45° and 90° with the horizontal and also strain gauges located at half the height of the panel.

The rupture pattern of the S1-2 panel was characterized by the appearance of a vertical fissure in the center of the panel that propagates in a direction parallel to the axis of application of the load to the vicinity of the nodal zone, as can be seen in Fig. 4.4.

It is noticed from the examination of Fig. 4.4 a typical rupture of traction being visible, additionally, the formation of compression wedges at the ends of the panel with localized crushing of the concrete. These compression cones (isosceles triangles) were also observed by the author in other tested panels and Fig. 4.5 illustrates this situation for panels S1-4, S2-8 and S3-10 which, although they have different characteristics of the S1-2 panel, exhibited similar rupture pattern.

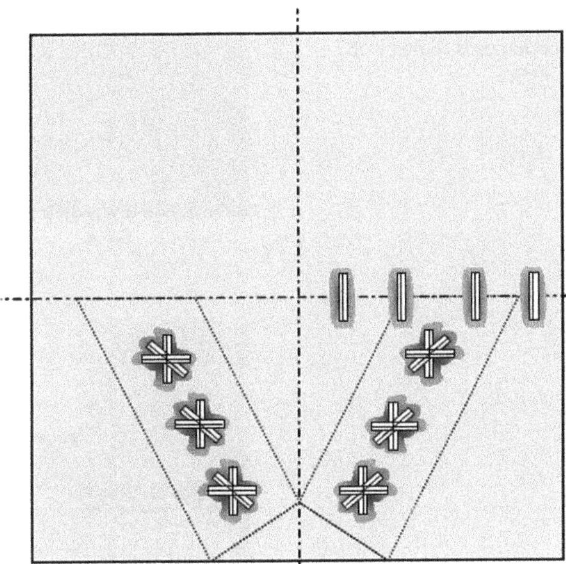

Fig. 4.3 Instrumentation of the S1-2 panel (Sankovich, 2003)

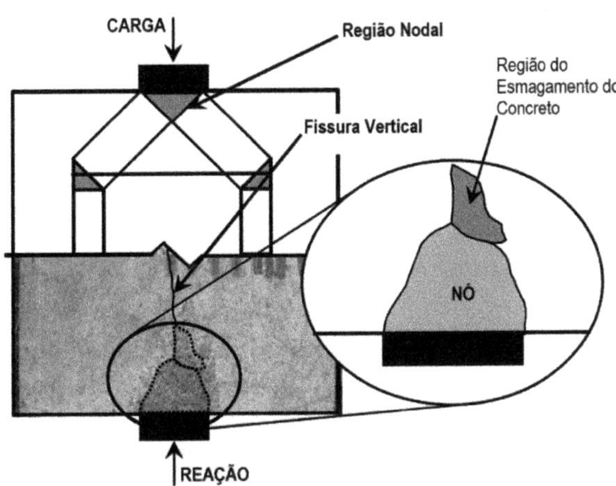

Fig. 4.4 Failure engine mechanism (Brow et al. 2006)

For the S1-2 panel, the first fissure load measured was 524 kN and the rupture load was 706.8 kN. The efficiency factor observed for this panel was 0.86 and the maximum tensile deformation measured was 0.0022.

Fig. 4.5 Rupture mechanisms (Sankovich 2003)

4.2.2 Panel S3-1

The typical geometry of the investigated elements consists of simple concrete panels with dimensions of 914.4 × 914.4 × 152.5 mm that were partially loaded using steel plate with a dimension of 304.8 × 152.5 × 50.8 mm until rupture, as illustrated in Fig. 4.6. Table 4.1 summarizes the data from the two panels used in this research

The instrumentation of this panel was performed as shown in Fig. 4.6 In it can be observed the strain gauge arranged at half the height of the panel.

The rupture pattern of panel S3-2 was identical to that observed in Panel S1-2 and the comments made there are valid here. For the S3-1 panel, the first fisura load measured was 608.5 kN and the rupture load was 873.2 kN. The efficiency factor observed for this panel was 0.65 and the maximum tensile deformation measured was 0.0014. Table 4.1 summarizes the information on the two panels under discussion, where the loads of first crack, the last load and the maximum tensile deformation at half the height of the panel can be observed.

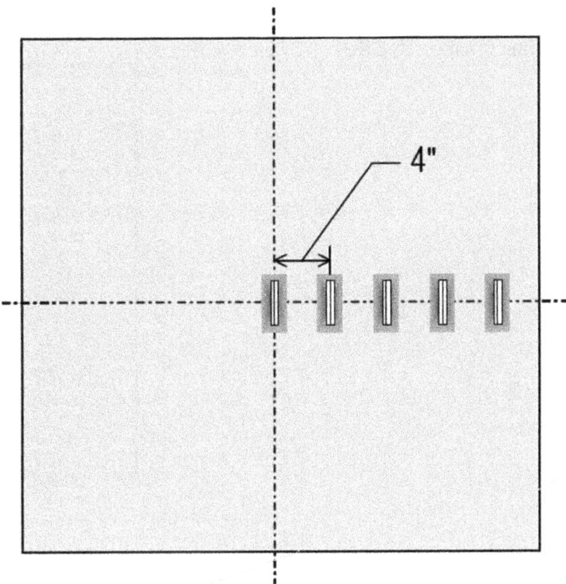

Fig. 4.6 Instrumentation of the S3-1 panel (Sankovich 2003)

4.2.3 Qualitative Information from Other Panels

In all panels investigated by Sankovich (2003), the failure mechanism was similar. The author explains that the nodal zone located near the loading application region has similar geometry of an isosceles triangle, where its height is half the width. It also highlights that when the fissure approaches the nodal zone, a change in its direction is observed with a decomposition of the same in two: one that leaned to the right and another to the left, bypassing the compression wedge, as shown in Fig. 4.7.

With regard to the deformations measured on the surface of the panels, Brown et al. (2006) points out that the first fissure loads varied between 37 and 81% of the rupture load. The average cracking load was 61% of the rupture load and the results are arranged within two standard deviations of the mean value.

According to Brown et al. (2006), for all the panels investigated the largest vertical deformations were measured in the central vertical line—loading application line—and were decreasing as it moves away from the center of the panel. The authors additionally reported traction deformations near the outer edge of the panel, in addition to important compressive deformation along the axis. Graph 4.1 shows the variation of vertical deformations along the distance to the center of the panel for four tested typologies, measured at the time of rupture, and illustrates well this effect that will influence the change in the configuration of the connecting rods.

Figure 4.7 shows the vertical deformations of panel S3-10 in two distinct loading moments—one in the formation of the first fissure and the other in the rupture.

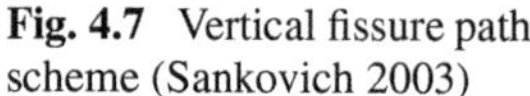

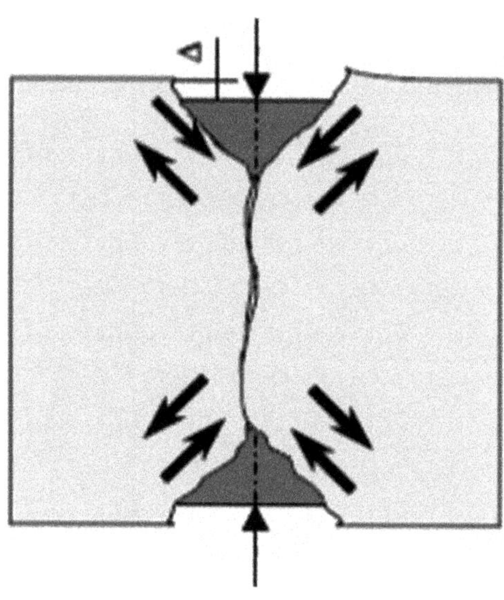

Fig. 4.7 Vertical fissure path scheme (Sankovich 2003)

References

Brown D, Sankovich CL, Bayrak O, Jirsa JO (2006) Behavior and efficiency of bottle shaped struts. ACI Struct J 103(3):348–355

Sanchez LFM, Drimalas T, Fournier B, Mitchell D, Bastien J (2018) Comprehensive damage assessment in concrete affected by different internal swelling reaction (ISR) mechanisms. Cem Concr Res 107:284–303

Sanchez LFM, Fournier B, Jolin M, Bastien J (2015) Evaluation of the Stiffness Damage Test (SDT) as a tool for assessing damage in concrete due to ASR: input parameters and variability of the test responses. Constr Build Mater 77:20–32

Sanchez LFM, Fournier B, Jolin M, Bastien J, Mitchell D (2016) Practical use of the Stiffness Damage Test (SDT) for assessing damage in concrete infrastructure affected by alkali-silica reaction. Constr Build Mater 125:1178–1188

Sanchez LFM, Fournier B, Jolin M, Mitchell D, Bastien J (2017) Overall assessment of Alkali-Aggregate Reaction (AAR) in concretes presenting different strengths and incorporating a wide range of reactive aggregate types and natures. Cem Concr Res 93:17–31

Sanchez LFM, Multon S, Sellier A, Cyr M, Fournier B, Jolin M (2014) Comparative study of a chemo–mechanical modeling for alkali silica reaction (ASR) with experimental evidences. Constr Build Mater 72:301–315

Sanders DH (1990) Design and behavior of anchorage zones in post-tensioned concrete Structures. Ph.D. Dissertation, The University of Texas at Austin, Austin, Texas, USA

Sankovich CL (2003) An explanation of the behavior of bottle-shaped struts using stress fields. MSc thesis, University of Texas, Austin, USA

Wollmann GP (1992) Anchorage zones in post-tensioned concrete structures. Ph.D. Dissertation, The University of Texas at at Austin, Austin, Texas, USA

Chapter 5
Numerical Simulation

5.1 Computational Model Description

The first step to modelling in ABAQUS software is the creation of the distinct parts that make up the model,—Module Parts. In this step, the specific geometry of the part to be modelled is defined. To do this, it is necessary to create the component, as shown in Fig. 5.1. In it, it is possible to observe the five settings necessary for the creation of the S1-2 panel—*Name, Modelling Space, Type e Base Feature.*

The plates for the application of loading and imposition of the boundary conditions were created as rigid shells. This option was adopted as a strategy to solve the problems of numerical singularity associated with the steel-surface plate contact of the concrete. The specific details of this step are illustrated in Fig. 5.2.

Figure 5.3 shows in addition to the panels (see Fig. 5.3b), the elements representing the top and base plates of each panel (see Fig. 5.3a).

In order to broaden the field of scientific observation of the research, a strategy was adopted to admit that the internal expansion reactions would be installed primarily in the bottle-shaped compression rod, defined from a linear elastic analysis of the panel. The decision to create a scenario like this, despite representing a rare event in the common situations, aims to provide information about the consequences of this rare event, but not impossible to occur, in the load capacity of the elements studied. This strategy proved justified and relevant because the results obtained are instigating, as will be presented and discussed in subsequent sections.

The area of the panel corresponding to the connecting rod was created using a Partition in the panel, as illustrated in Fig. 5.4. Figure 5.5 displays the result of the strategy used for panel S3-1.

A. C. Azevedo et al., *Concrete Structures Deteriorated by Delayed Ettringite Formation and Alkali-Silica Reactions*, Building Pathology and Rehabilitation 24,
https://doi.org/10.1007/978-3-031-12267-5_5

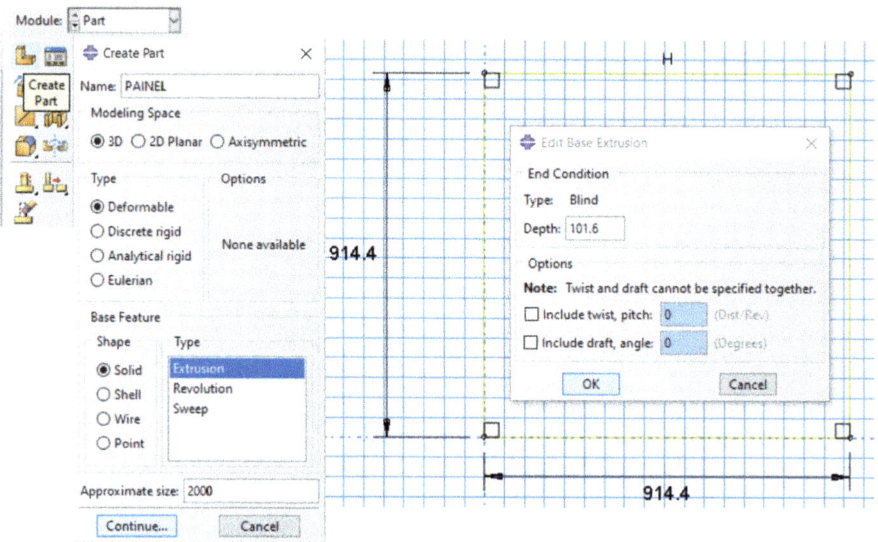

Fig. 5.1 Panel S1-2 creation

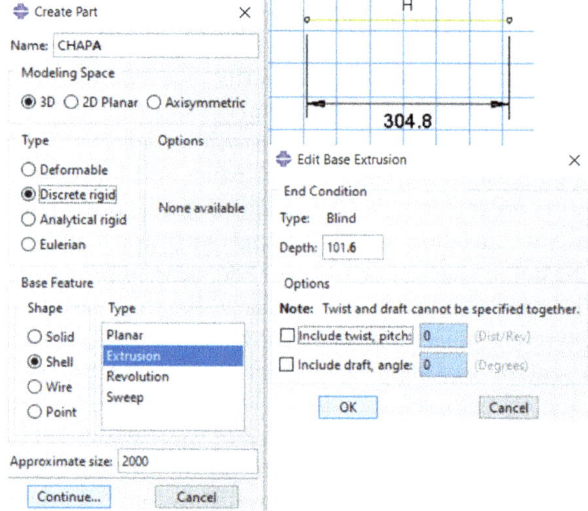

Fig. 5.2 Creation of the S1-2 plate

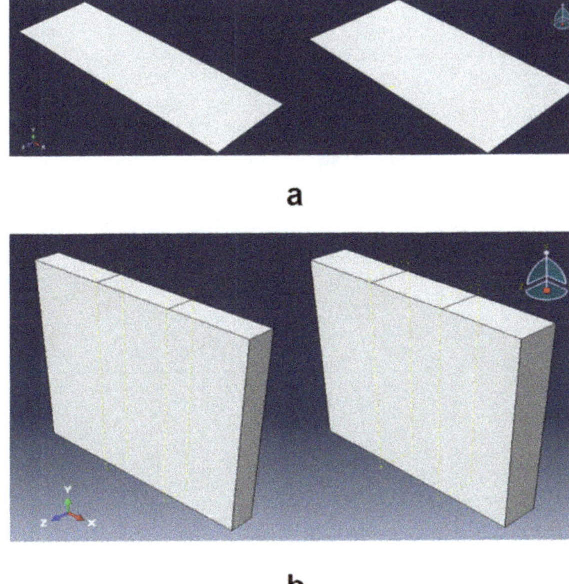

Fig. 5.3 **a** Top plates S1-2 and S3-1, **b** S1-2 and S3-1

5.2 Computational Model

The finite element used to model of all panels studied was the C3D8R. The steel plates were modelled with the Discrete Rigid Element—R3D4 family. A structured finite element mesh composed exclusively of cubic elements of edges equal to 50.8 mm was used. This decision was taken after the study of several other modelling possibilities and was the one that presented a better computational effort/work ratio of analysis and interpretation of results.

The resulting finite element mesh for panels S1-2 and S3-1 are shown in Figs. 5.6 and 5.7, respectively. The mesh of the plates is shown in Fig. 5.8. For the modelling of this panel with the upper and lower steel plates, 1,008 elements and 1,500 nodes were used, totalling 4,500 degrees of freedom. It was also considered a contact interaction between the steel plates and the panel surface with a coefficient of friction of 0.3.

The applied load was an uniformly distributed compression stress on the contact surface of steel plate/concrete panel, corresponding to the value of the rupture load of the experimental tests (Sanchez et al. 2014). The boundary conditions of the top plate were set so that only vertical displacements were enabled. This boundary condition allows the rigid body displacement of the plate in the direction of loading. The same procedures used to impose the boundary conditions on the upper plate were used on the lower plate, but here all three-translation degree of freedom were restrained.

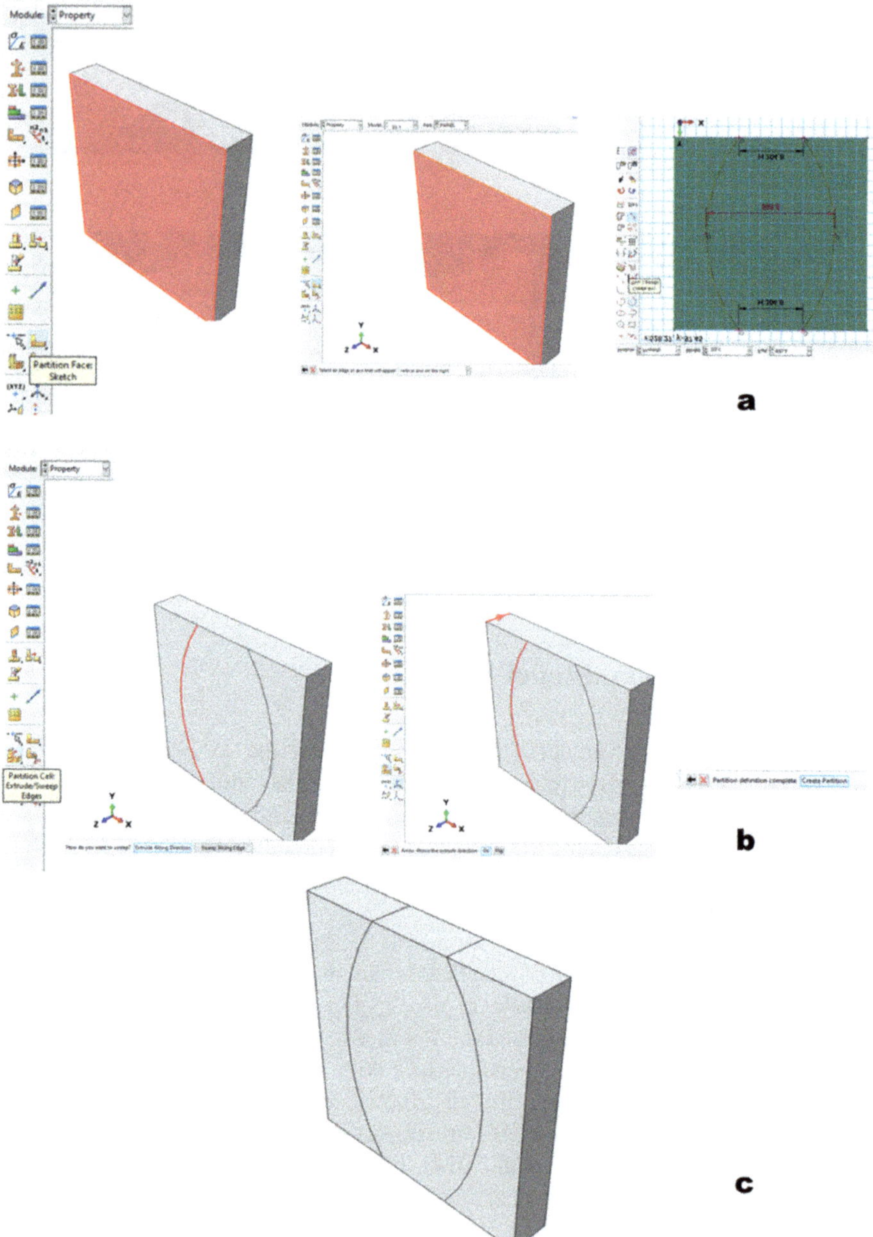

Fig. 5.4 Partition for delimitation of the connecting rod

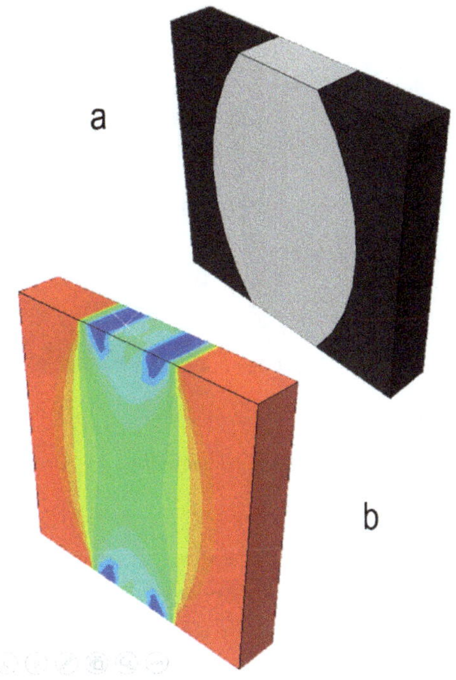

Fig. 5.5 S1-2 with the strut from linear elastic analysis of the panel

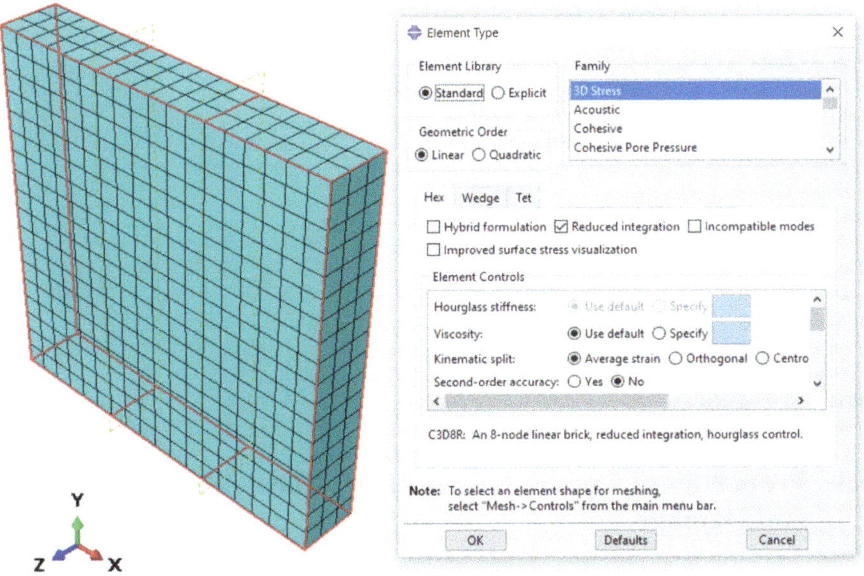

Fig. 5.6 Finite element mesh of S1-2 panel

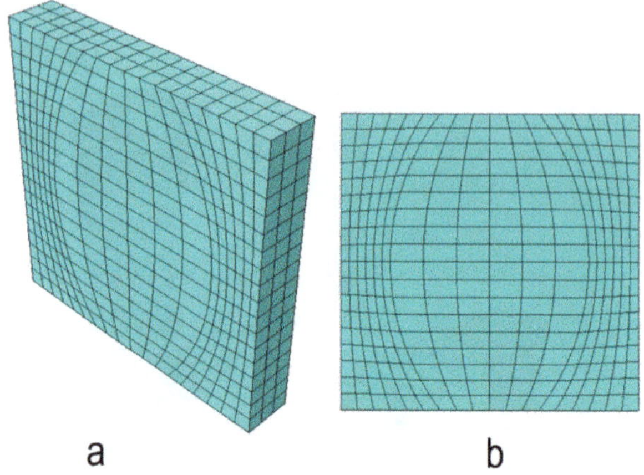

Fig. 5.7 S3-1 finite element mesh with detached connecting rod **a** perspective, **b** front

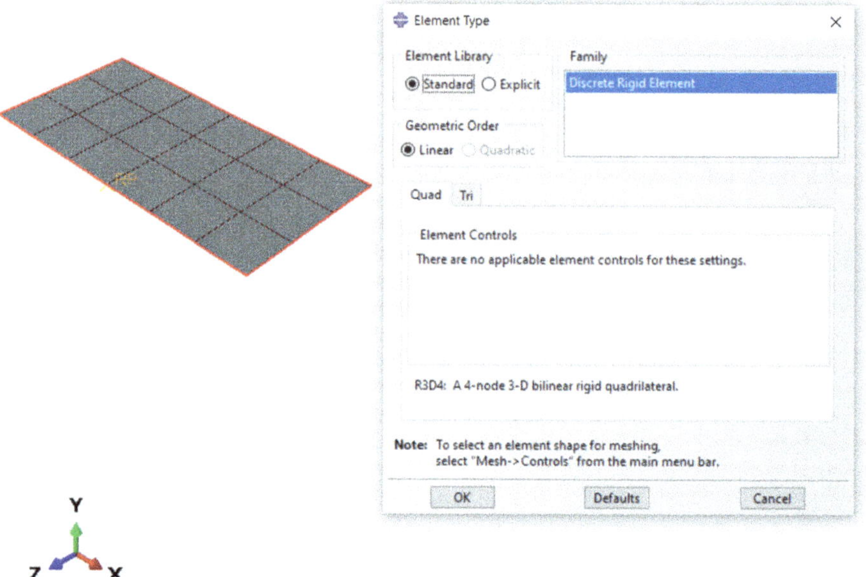

Fig. 5.8 Finite element mesh of the loading plates

5.3 Properties of Concrete—Elasticity, Plasticity and Damage

To define the properties of the concrete material in ABQUS, the definition of values associated with the elastic-linear and nonlinear behaviour of the material is necessary. In the modelling of the nonlinear behaviour, we used the data obtained from the CDP

Table 5.1 Input concrete properties for CDPM in ABAQUS (S1-2)

Elasticity		Plasticity	
Compressive strength, f′c (MPa)	26.41	Dilation angle	5
		Eccentricity	0.1
Elasticity model (MPa)	28,564.32	fb0/fc0	1.16
		K	2/3
Poisson's ratio	0.2	Viscosity	0.01
Damage			
Compressive behaviour		Compressive damage	
Yield stress (MPa)	Inelastic strain	Damage parameter	Inelastic strain
34.41	0	0	0
20.50	0.00561	0.57	0.00561
13.62	0.00792	0.73	0.00792
8.22	0.0109	0.86	0.0109
1.92	0.0228	0.99	0.0228
Tensile behaviour		Tensile damage	
Yield stress (MPa)	Cracking strain	Yield stress (MPa)	Cracking strain
2.67	0	0	0
2.02	0.0002	0.20	0.0002
1.76	0.0003	0.3	0.0003
1.35	0.0005	0.46	0.0005
1.06	0.0007	0.59	0.0007
0.72	0.0012	0.81	0.0012
0.03	0.0046	0.999	0.0046

routine described in Sect. 3.4. Tables 5.1 and 5.2 summarize the data that were used, respectively, in the modelling of panels S1-2 and S3-1.

5.4 Step Definition

To run a simulation in ABAQUS, you must define an increment—called *Step*. This parameter defines the division of the loading history into steps. In the analysis performed, two general steps were defined, i.e.:

- An initial step, Initial Step, automatically generated in the creation of the model, from which the boundary conditions are defined;
- One step of analysis, *Static Step*, static analysis, where loads/displacements were imposed.

Table 5.2 Input concrete properties for CDPM in ABAQUS (S3-1)

Elasticity		Plasticity	
Compressive strength, f′c (MPa)	28.96	Dilation angle	5
		Eccentricity	0.1
Elasticity model (MPa)	29,446.22	fb0/fc0	1.16
		K	2/3
Poisson's ratio	0.2	Viscosity	0.0046
Damage			
Compressive behaviour		Compressive damage	
Yield stress (MPa)	Inelastic strain	Damage parameter	Inelastic strain
36.96	0	0	0
21.58	0.00786	0.54	0.00786
14.19	0.0113	0.72	0.0113
8.02	0.0163	0.87	0.0163
1.79	0.0357	0.99	0.0357
Tensile behaviour		Tensile damage	
Yield stress (MPa)	Cracking strain	Yield stress (MPa)	Cracking strain
2.84	0	0	0
2.14	0.0002	0.20	0.0002
1.87	0.0003	0.3	0.0003
1.43	0.0005	0.46	0.0005
1.13	0.0007	0.59	0.0007
0.71	0.0012	0.81	0.0012
0.04	0.0046	0.999	0.0046

Figure 5.9 shows the definitions made in this step.

5.5 Assembly

The *Assembly* module is used to join all the created parts and sort them in the appropriate positions to the model under study. Figure 5.10 shows the association of parts performed in one of the models studied.

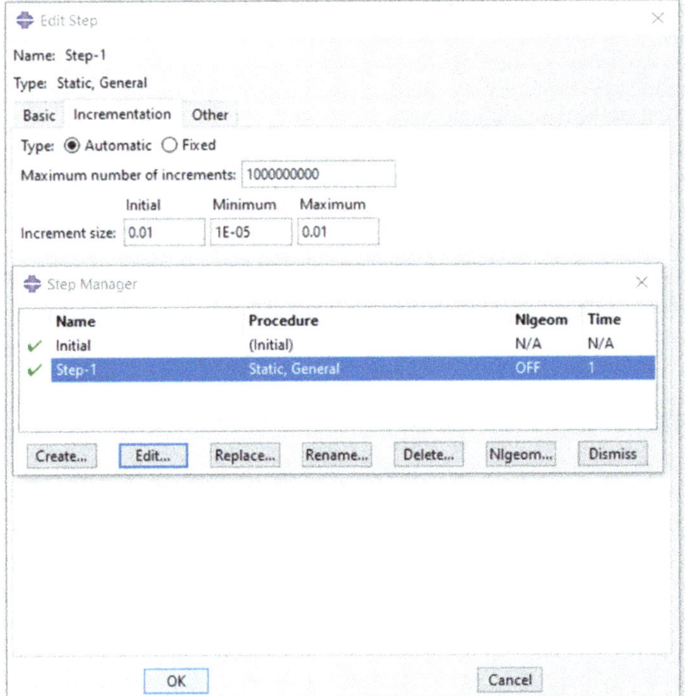

Fig. 5.9 Definition of the step

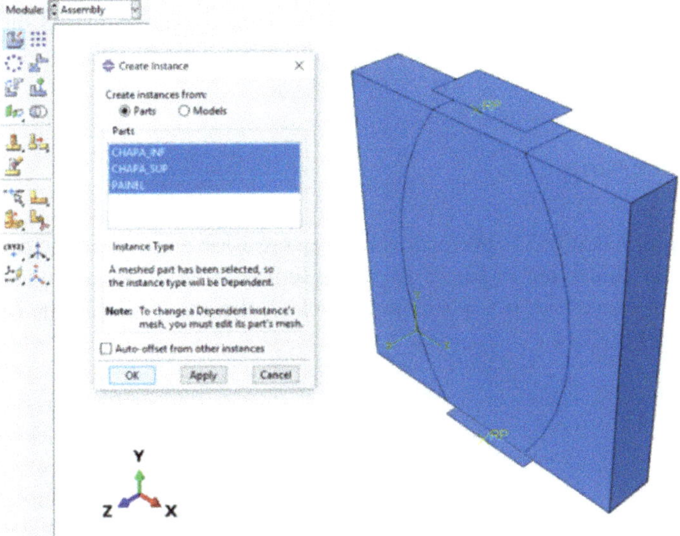

Fig. 5.10 Association of the components of a typical panel

5.6 Definition of Interactions in the Model

ABAQUS provides a module to define and manage interactions in the model. Interactions define actions or dependencies between two or more points or surfaces. A contact interaction was generated, contact interactions, between the surfaces of the plates and the panels. This interaction allows the transfer of normal stresses and displacements in the steel plate to the concrete surface. The coefficient of friction used was 0.3 which is commonly applied for this type of contact. The plates have been set to Master Surface and the panel surface in the loading application region as Slave Surface. Figure 5.11 shows the details of the settings made.

5.7 Definition of the Boundary Conditions

The Load module is used to define loads, forces, initial temperatures, displacements and boundary conditions. The creation of the Steps conditions both the definition of supports and loads. In this research, we chose to apply a concentrated force at a reference point in the upper steel plate and the condition adopted as a rigid plate allowed the uniformity of the displacements generated by the load.

The loads used in the simulations were those reported at the rupture by Brown et al. (2006) for the panels studied (see Table 4.1). The contour condition of the upper plate was defined in the *Step Initial* so that only the offset in the vertical direction was enabled. This setting allows the displacement of the board's rigid body in the direction of loading. The same procedures used to impose the boundary conditions on the upper plate were used in the lower plate, with translation restriction in all three directions. Figures 5.12 and 5.13 illustrate the procedures discussed.

5.8 Processing

Processing in ABAQUS is performed by creating a Job and setting the properties for execution, as illustrated in Fig. 5.14. After setting these properties, the template is submitted to execution. It is possible to track all the processing in the Monitor tab, as shown in Fig. 5.14.

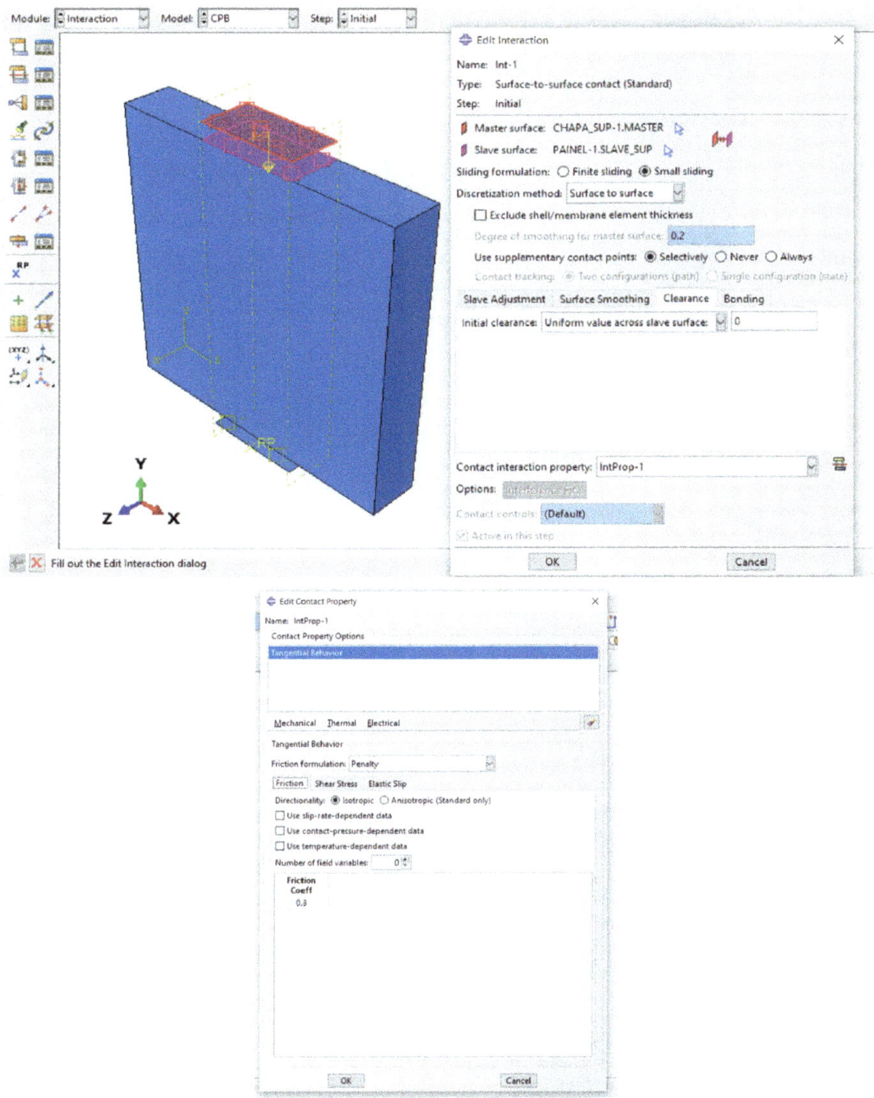

Fig. 5.11 Contact interaction between plates and panel

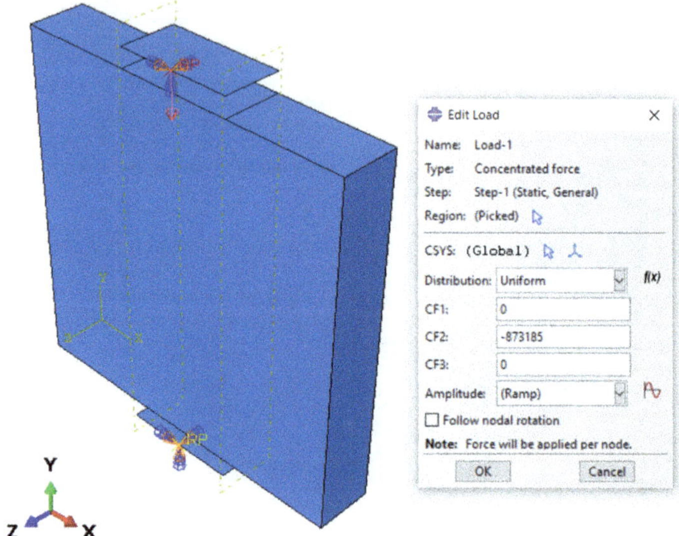

Fig. 5.12 Loading application

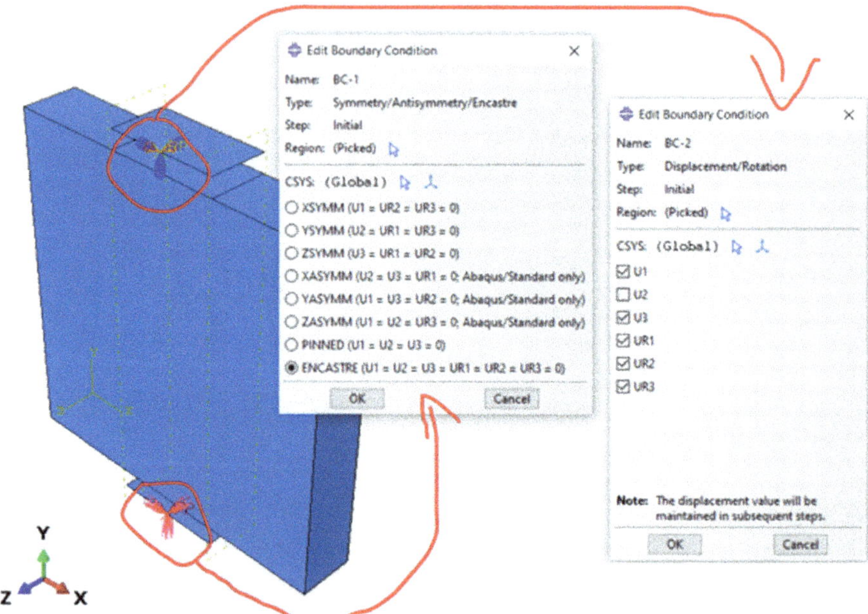

Fig. 5.13 Boundary conditions

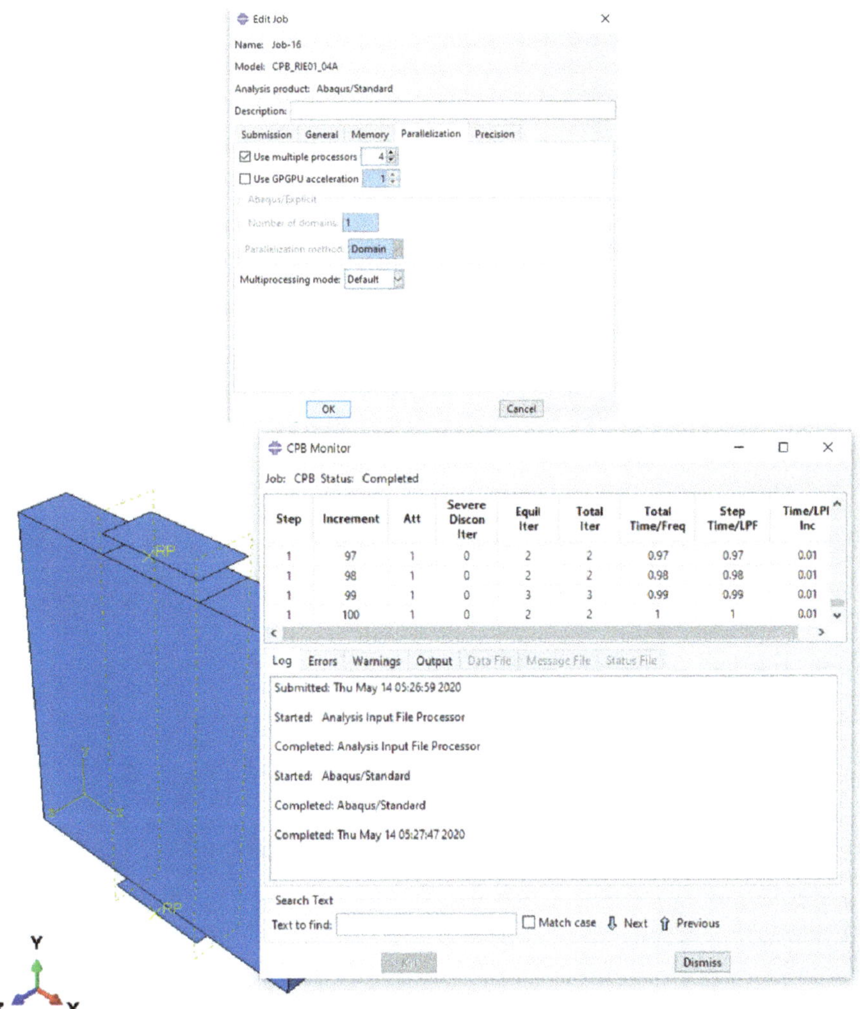

Fig. 5.14 Definition of the job and processing

References

Brown D, Sankovich CL, Bayrak O, Jirsa JO (2006) Behavior and efficiency of bottle shaped struts. ACI Struct J 103(3):348–355

Sanchez LFM, Multon S, Sellier A, Cyr M, Fournier B, Jolin M (2014) Comparative study of a chemo–mechanical modeling for alkali silica reaction (ASR) with experimental evidences. Constr Build Mater 72(2014):301–315

Chapter 6
Numerical Results and Discussion

6.1 FEM Validation

The following qualitative and quantitative comparisons are made between the numerical models developed and the experimental results presented in Chap. 4.

6.1.1 Qualitative Comparisons

A plastic tensile strain of the model of S1-2 panel output is presented in Fig. 6.1 close to the failure view of the panel observed in lab tests. In this figure, it is possible to observe a good correlation between the profiles of plastic tensile strains with those observed in the lab tests. This fact highlights that the numerical model developed was able to adequately describe the mechanisms involved in the rupture of the panel understudy—formation of the compression cone at the top and base of the panel and changing the orientation of the plastic tensile strains in the vicinity of the compression cone.

The nodal region, both in the numerical and lab test models, presents the shape similar to that of an isosceles triangle with its base being two times its height. In this region, the stresses are mostly compression and, in its vicinity, can be observed the crushing of concrete (see Fig. 6.5b).

The tensile damage profile of the numerical model is displayed in Fig. 6.2 close to the S1-2 panel view at failure. In this Figure, it is possible to observe that the numerical model was able to describe the damage that occurred on the lateral faces of the panel. Sankovich (2003) and Brown et al. (2006) reported the existence of damage in regions close to the compression edges at the top and base of the panel, a situation also described by the numerical model developed—damage of approximately 40%. Maximum tensile damage has reached to 100% in the central region of the panel, at failure.

A. C. Azevedo et al., *Concrete Structures Deteriorated by Delayed Ettringite Formation and Alkali-Silica Reactions*, Building Pathology and Rehabilitation 24,
https://doi.org/10.1007/978-3-031-12267-5_6

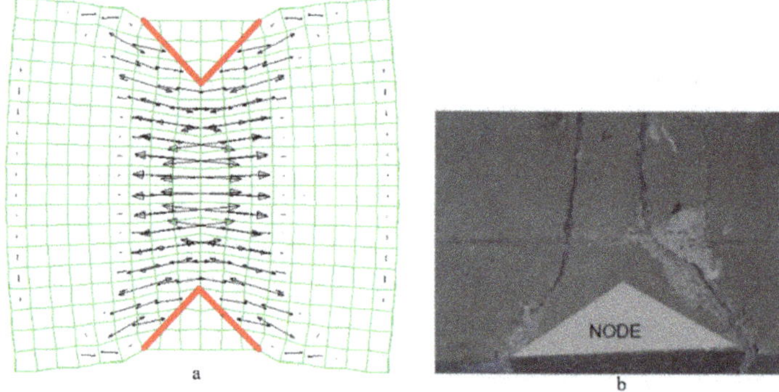

Fig. 6.1 Experimental S1-2. **a** Numerical and **b** Plastic tensile strains

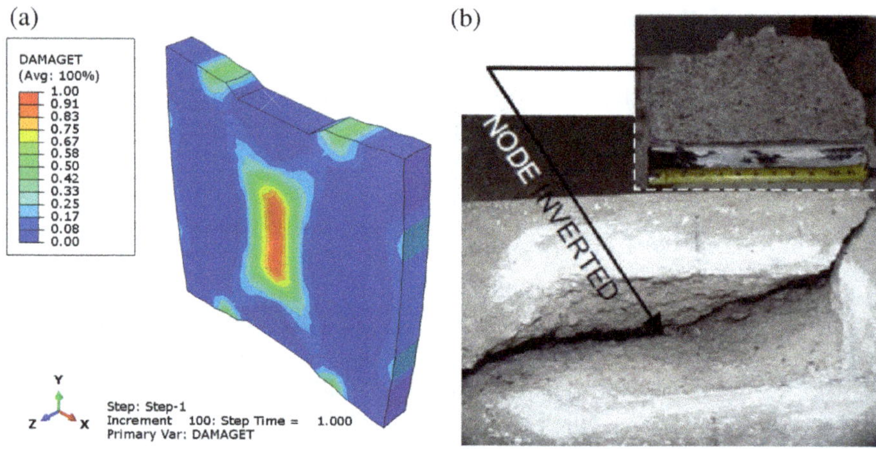

Fig. 6.2 Tensile damage of the S1-2 panel. **a** Numerical model and **b** Lab test

The compression damage in the numerical models was less pronounced than the tensile damage, an aspect also verified in the lab tests. The magnitude of the damage in compression of the numerical model was approximately 71%. Despite this fact, the profile of compression damage is relevant to validate the location of the compression crushing of the concrete in the numerical models. Figure 6.3 shows the profile of compression damage, where the concentration of damage can be observed in the vicinity of the nodal region, consistently with crash regions of concrete observed in lab tests. It was possible to capture a strut narrowing with the increase of vertical load, during the loading process. Figure 6.4 exhibits this behaviour highlighting the tension and compression fields in the numerical model of S1-2 panel, which is quite similar to those observed in lab test.

The numerical model also captured the occurrence of vertical tensile stresses that are responsible for the narrowing of the strut observed in the lab tests. This fact

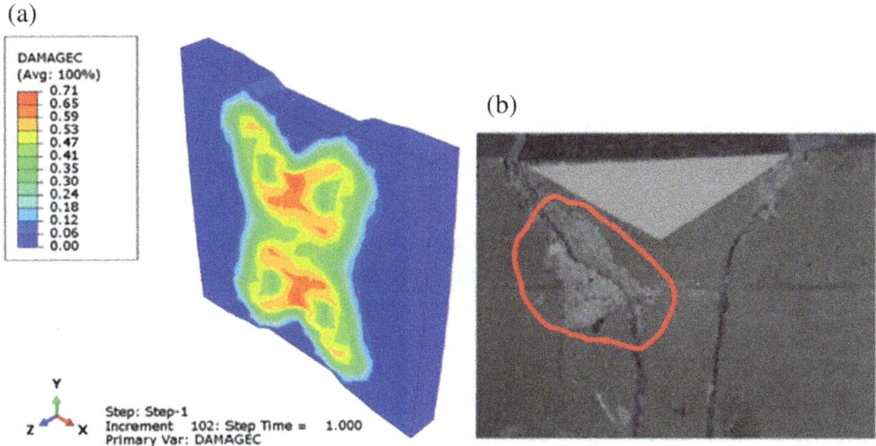

Fig. 6.3 Compressive damage in S1-2 panel. **a** Numerical model and **b** Lab tests

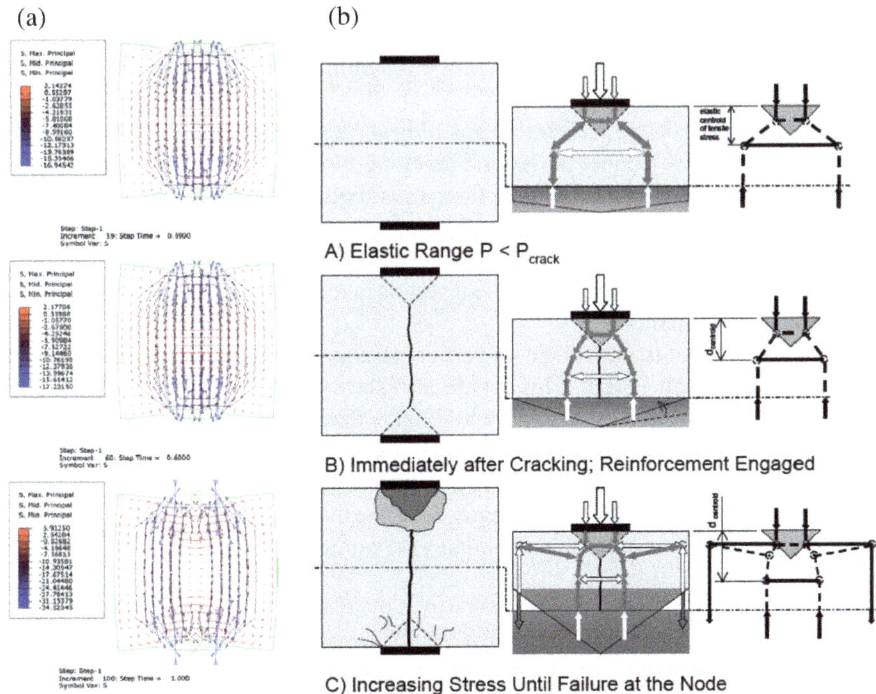

Fig. 6.4 Evolution of STM for different loading levels. **a** Numerical model and **b** Lab tests

demonstrates that the numerical model was efficient in capturing the phenomenon discussed.

6.1.2 *Quantitative Comparisons*

Table 6.1 summarizes the lab tests and numerical results for the panels investigated. In this Figure, the first cracking load and the maximum horizontal tensile strains are presented for the numerical model and lab tests. As it can be observed, the numerical model was able to predict the first cracking load of the investigated panels with acceptable precision, considering the complexity of the studied phenomenon. The maximum error in the first cracking load and the maximum tensile strain was −19.1% and −14.30%, respectively.

The distribution of vertical deformations along the width of the panels was measured by strain gauges distributed at their half height from the center to its outer edge. When performing the tests Sankovich found the symmetry in the distribution of vertical deformations, and, because of this fact, decided to instrument only the right half of the panels. Figure 6.5 present this symmetry in the numerical models in the two panels investigated.

Two verifications were performed regarding vertical deformations: one comparing the results of the S3-1 panel at the first cracking formation and the other at failure with data of the S3-10 panel. This option is justified by the similarity in the instrumentation of these two panels. Figure 6.6 presents this comparison. The behaviour of the numerical model for the first cracking and failure loads was similar to those observed in lab tests—maximum vertical strains in the center of the panel decreasing to the edges of the panel.

Figure 6.7 shows the envelope of vertical strains measured at half the height of the panels along their length. This Figure also shows the strains obtained in the lab tests. The results showed in this figure highlights that the responses of the numerical panels investigated were located inside the experimental envelope.

Given the results presented in this section, one can conclude that the numerical models developed were able to efficiently describe the experimental behaviour of the panels investigated, an aspect that validates the numerical models.

Table 6.1 Comparison of numerical and lab tests results

Panel	First cracking load, (kN)		Error (%)	Maximum tensile strain, (‰)		Error (%)
	Num	Exp		Num	Exp	
S1-2	424.1	524.0	−19.1	0.0025	0.0022	−13.64
S3-1	681.1	608.5	11.93	0.0012	0.0014	−14.3

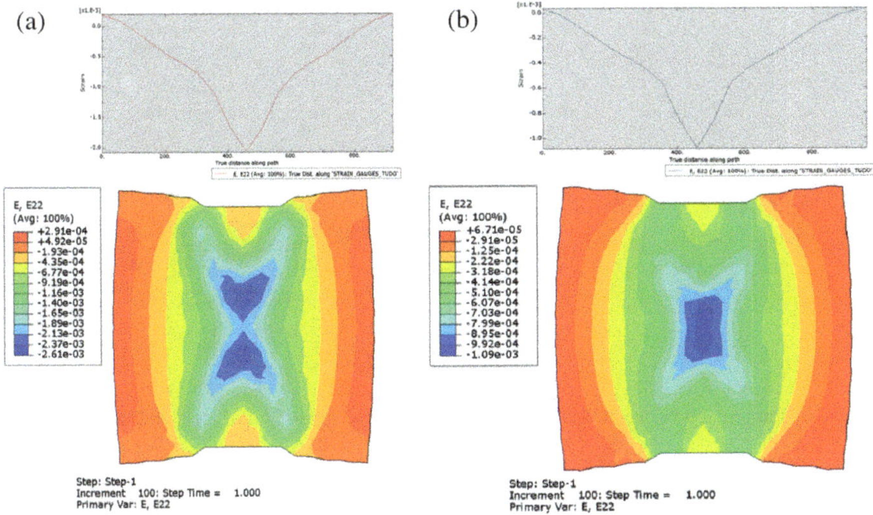

Fig. 6.5 **a** Symmetry of vertical deformations of S1-2 and **b** Symmetry of vertical deformations of S3-1

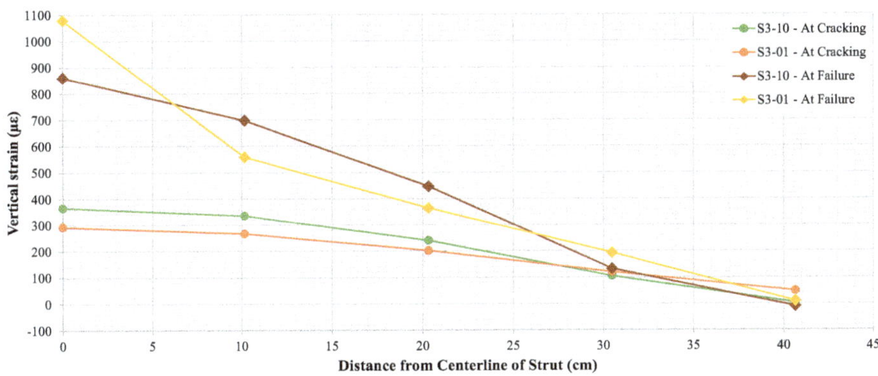

Fig. 6.6 Comparison between vertical strains at first cracking and failure of S3-10 and S3-1 panels

6.2 Panel S3-1 Deteriorated by Alkali-Silica Reaction (ASR)

Figure 6.8 shows the evolution of load and tensile damage with the expansion level at the first crack formation. It can be observed that the panel without internal expansion, hereinafter referred to as intact panel, presented a first crack load of approximately 681 kN and for the expansion level of 0.04% the first crack load was about 297 kN. This means that for a relatively low level of expansion the first crack load decreasing

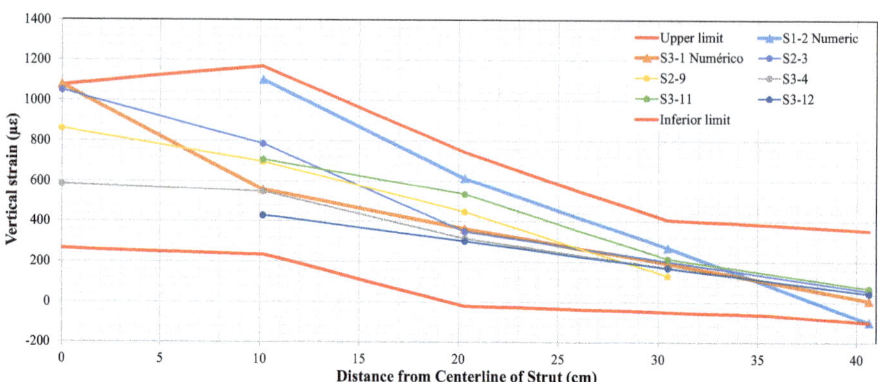

Fig. 6.7 Comparison between vertical strains at failure for numerical models and lab tests results

was close to 57%. For the expansion level of 0.30% the first crack load was 175 kN—a reduction of 74.3%. This highlights the important effect of the internal expansion reactions on the overall behaviour of the panel.

When one observes the damage, intact panel exhibited damage of approximately 0.06% at first crack formation and at the expansion level of 0.30% showed a maximum damage at first crack of approximately 0.20%—an increase of approximately 230%.

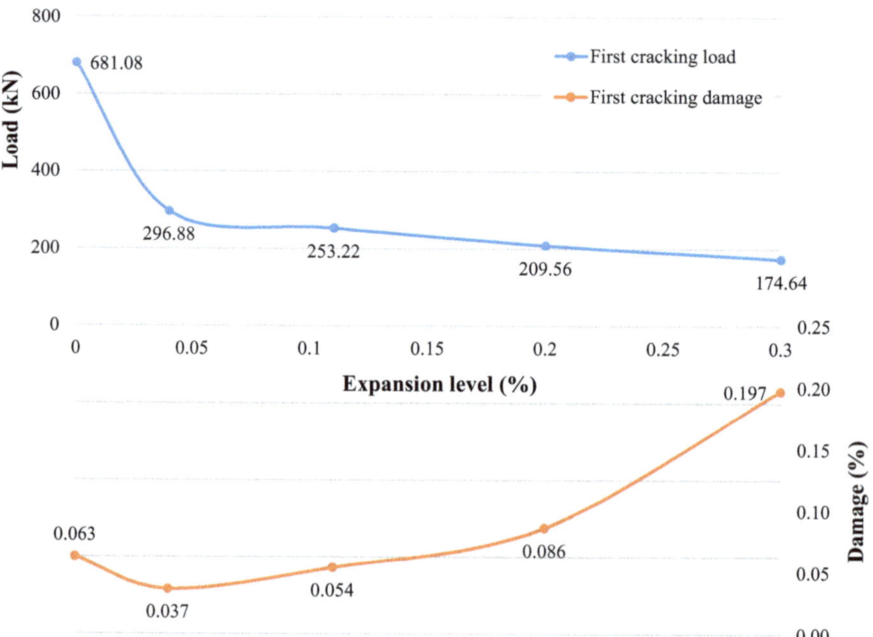

Fig. 6.8 First crack load and damage for each expansion level

This significant increase shows an important reduction in panel stiffness caused by the internal expansion reaction.

Figure 6.9 shows the evolution of the load and damage with the expansion level at failure. The failure load of the intact panel was 873 kN and, for the expansion level of 0.04%, its value reached 489.77 kN—a reduction of 44%. The failure load for the expansion level of 0.30% was even lower—281.86 kN—, configuring a reduction of approximately 70%. The damage at failure for the intact panel was 74.92% and for the expansion level of 0.30% its value was 96.33%—an increase of 28.58%.

It is also possible to observe in Figs. 6.8 and 6.9 that, for relatively discrete expansion levels, an important reduction is already observed, both at first crack and at failure loads. Additionally, it was observed that the effect of discrete expansions was more pronounced at first crack load. For increasing expansion levels, reductions in first crack and failure loads were less pronounced. This fact points to the importance of knowing the levels of expansion existing in concrete parts affected by expansive internal reactions in order to evaluate the need for retrofitting or rehabilitation works.

Figure 6.10 displays the tensile damage profile at the break for the 0.04% expansion levels for the healthy panel, the affected panel and the partially affected panel in the connecting area. It can be observed that the expansive reaction increased the range of damage along the panel, especially in the bottle-shaped connecting rod region. The maximum damage reached the value of 98% in the central region of

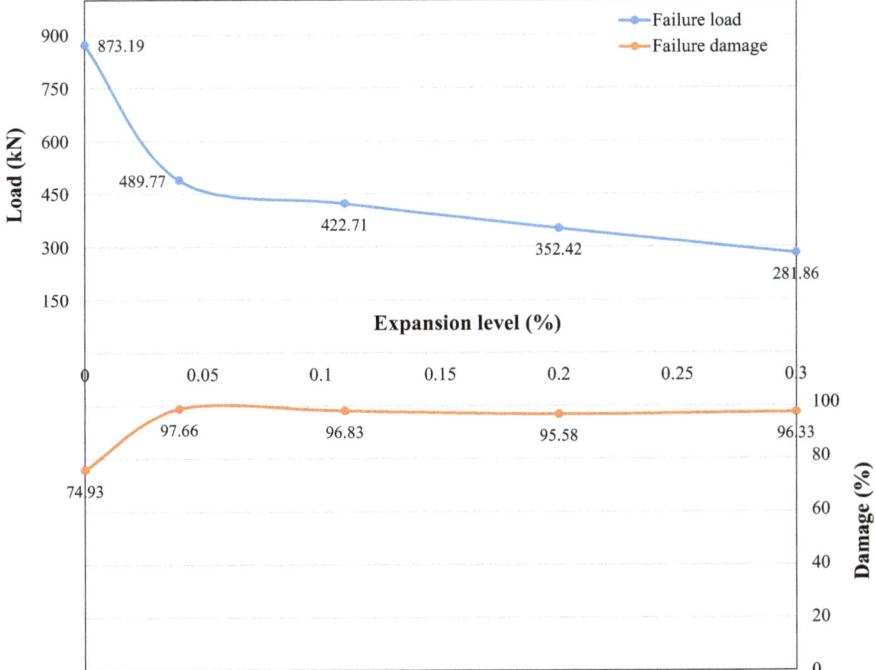

Fig. 6.9 Failure load and damage for each expansion level

the affected panels. It was also observed the appearance of traction damage on the sides of the loading zone of the panels, compatible with the experimental observations. This damage was more pronounced when only the connecting rod region was affected with the effects of the expansion reaction.

For the expansion level of 0.04%, the fully affected panel presented a load of 514 kN—5% higher than that corresponding to the panel affected by expansion reactions exclusively in the connecting rod.

Figure 6.10c (panel affected exclusively in the connecting rod region) shows that the damage propagates vertically in the direction of the loading axis, reaching the top of the panel, whereas in Fig. 6.10b (fully affected panel) the damage is primarily concentrated in the central part of the panel. It was still possible to observe damage to the lateral vertical faces of the fully affected panel.

The failure damage profile for expansion levels of 0.30% is shown in Fig. 6.11. For an expansion of 0.30%, the fully affected panel exhibited a tensile failure damage profile at the most distributed in the centre of the panel region. For this panel, damages above 40% were observed throughout the central region. The damage profile on the panel partially affected was more concentrated damage on the peripheries of the strut, adjacent to the compression cone.

The fully affected panels and the panels affected exclusively in the struts presented practically equal load capacity at failure, but with a different distribution of tensile damage profile. The explanation for this fact is that the mechanism of failure of the panels is complex. In fact, the expected failure mode is characterized by the appearance of an approximately vertical tensile crack that gradually increases in opening until a non-ductile failure characterized by crushing the compressive concrete in the vicinity of the nodal region. This last mode of failure was adequately captured by CDPM which showed plastic compression strains in this region, which numerically indicate the beginning of the concrete crushing process.

Figure 6.12 displays the tensile plastic deformations at the rupture to the expansion level of 0.04% for the healthy panel, for the fully affected panel and for the affected panel exclusively in the connecting rod. It can be observed that the profile of the plastic tensile deformations of the intact panel and the integrated panel are similar, although the maximum values of the deformations are different—1.28‰ and 2.27‰, respectively. However, when it is compared these two hypotheses with that of the panel affected exclusively in the connecting rod region, there are changes in the distribution profile of the deformations and at their maximum value—2.94‰. In the latter case, the most significant cracking process is located at the contour of the nodal region.

For an expansion level of 0.30% the profile of plastic tensile strains is different for the three hypotheses studied, as shown in Fig. 6.13, with the maximum values of 1.28‰, 2.05‰ and 2.06‰, respectively. The differences observed in the profile of plastic tensile strains for the studied expansion levels of the fully affected panel and the affected panel exclusively in the struts points to the importance in choosing the mechanism of allocation of the effects of the expansion in the concrete element

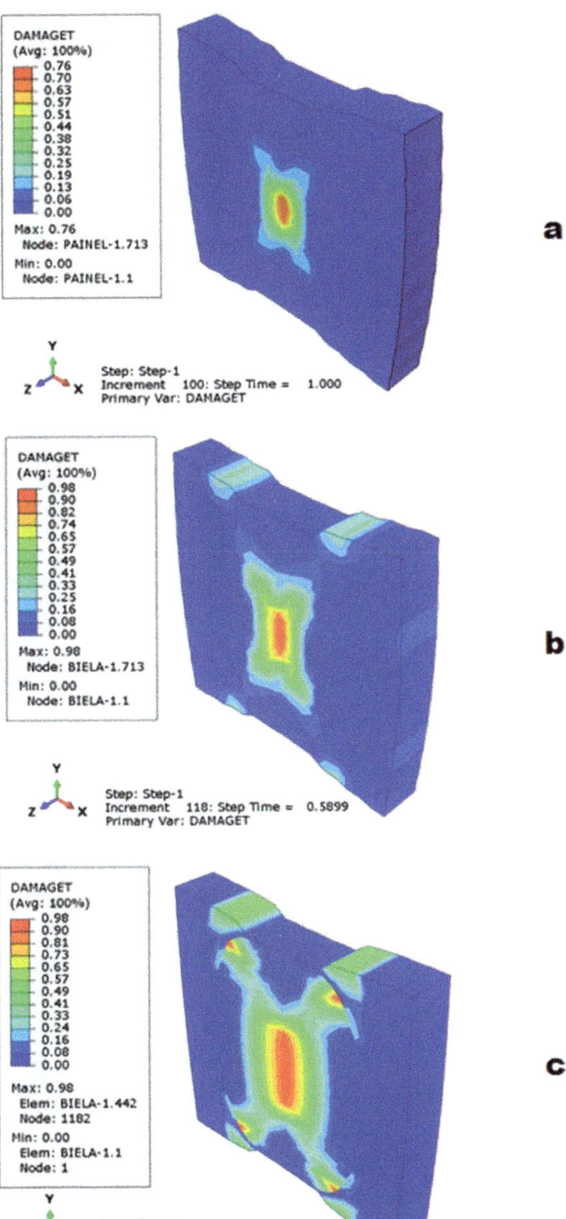

Fig. 6.10 Panel tensile damage for expansion of 0.40%. **a** Unaffected. **b** Fully affected. **c** Partially affected

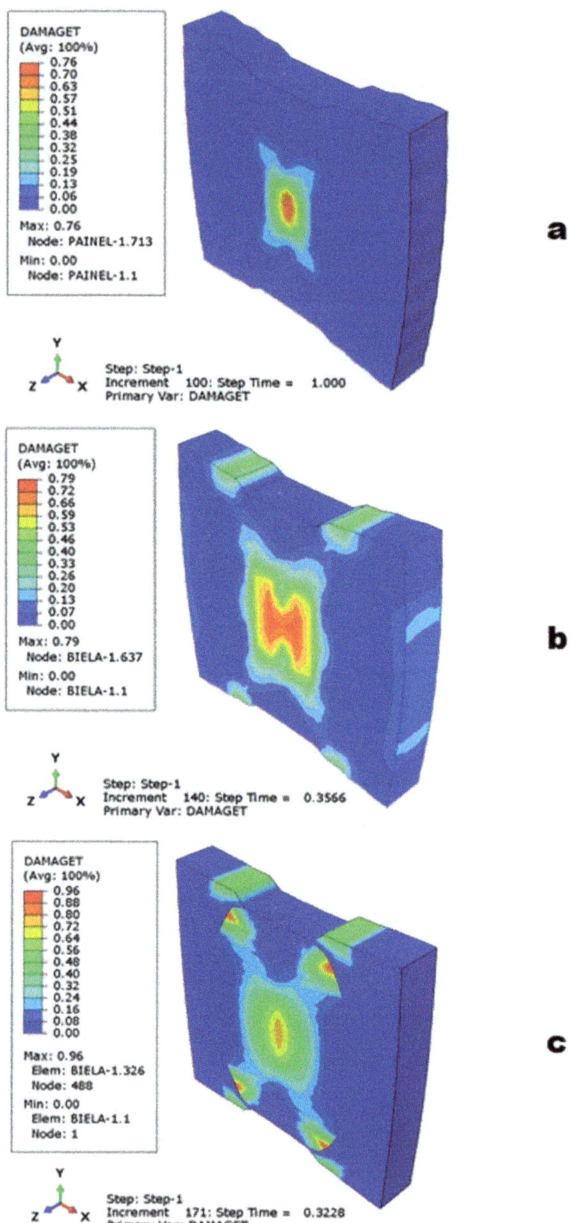

Fig. 6.11 Panel tensile damage for expansion of 0.30%. **a** Unaffected. **b** Fully affected. **c** Partially affected

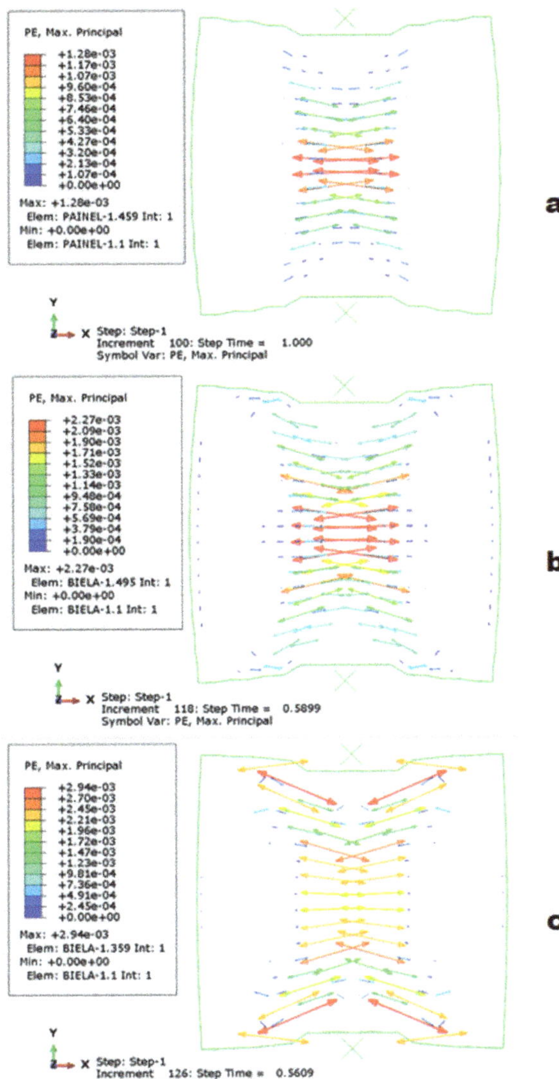

Fig. 6.12 Plastic tensile strains for expansion of 0.40%. **a** Unaffected. **b** Fully affected. **c** Partially affected

studied. This result shows that the strategy of affecting exclusively the strut is important. For the authors, the choice to affect the strut exclusively is more adequate because of the load support mechanism involved in the concrete element investigated.

The profile of stresses and deformations on the horizontal axis for the expansion level of 0.04% for the healthy panel, the panel integrally affected and for the affected panel exclusively in the connecting rod region is illustrated in Fig. 6.14.

Total tensile deformations are higher in panels with internal expansion reaction. However, the stresses on these panels are lower, indicating the influence of the reduction in tensile strength.

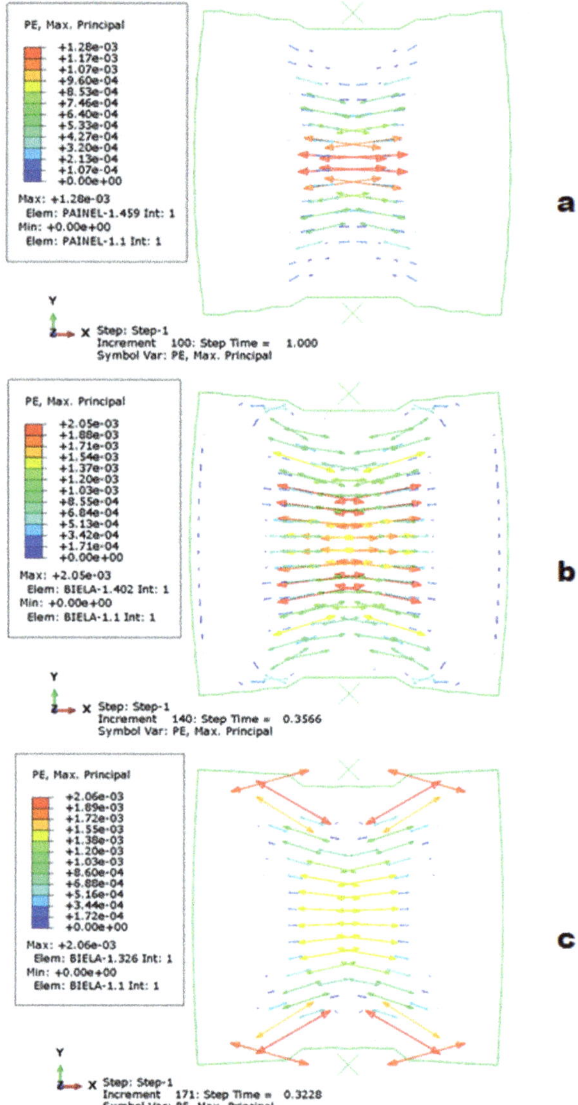

Fig. 6.13 Plastic tensile strains for expansion of 0.30%. **a** Unaffected. **b** Fully affected. **c** Partially affected

Total tensile deformation is lower on the partially affected panel compared to the fully affected panel. This fact evidences the confinement of the connecting rod caused by the unaffected portion of the panel. There is a higher compression stress component and very close to the contact between the connecting rod and the rest of the panel. This explains the strong inclination of the traction components and the change of the configuration in the voltage distribution. Figure 6.15 displays the stress and deformation profile on the horizontal axis for the expansion level of 0.20%.

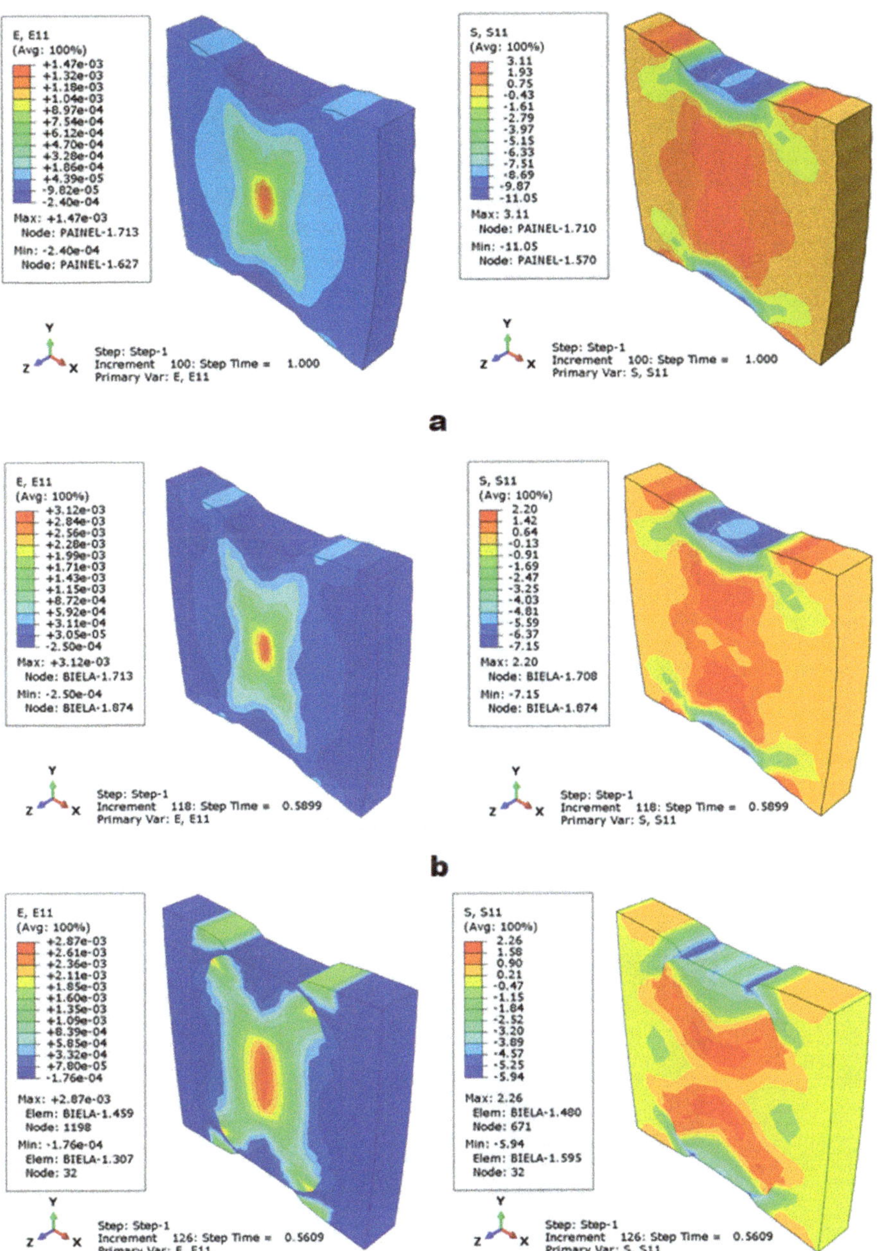

Fig. 6.14 Total deformation and x-axis voltage for expansion of 0.04%. **a** Unaffected. **b** Totally affected. **c** Partially affected

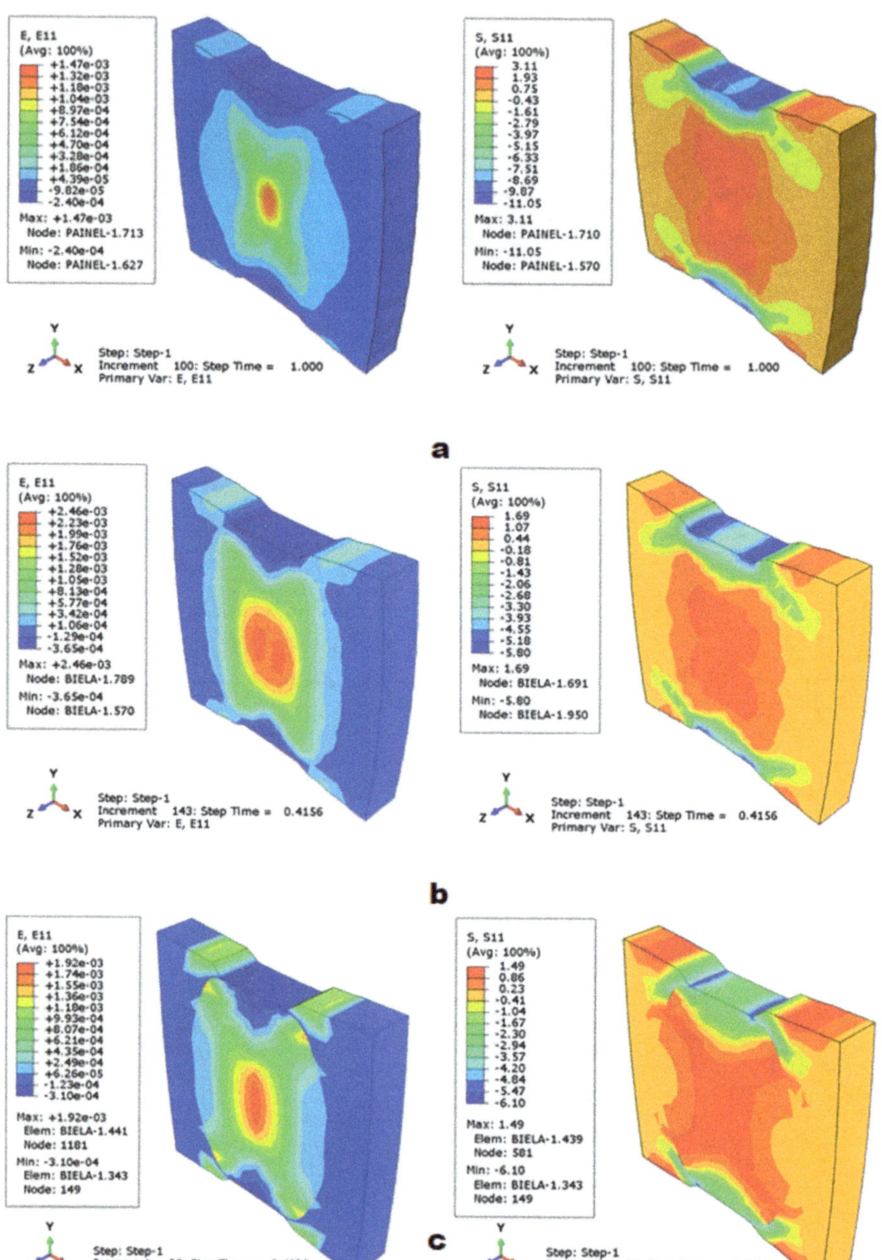

Fig. 6.15 Total deformation and x-axis voltage for expansion of 0.20%. **a** Unaffected. **b** Totally affected. **c** Partially affected

For the expansion level of 0.20%, the total tensile deformation of the partially affected panel is concentrated in the center propagates to the region of application of the load. At this stage of reductions in mechanical properties, the compression cone decreases.

In the fully affected panel the maximum tensile deformation ceases to be in the center and propagates to 5 cm on the right and left. This behaviour is consistent with that observed in the plastic tensile deformations of the previous figures. The maximum horizontal stress of panels with internal expansion ration is approximately 50% lower than that recorded on the panel without expansion. The reduction in minimum voltage follows approximately the same percentage.

Figure 6.16 shows the profile of horizontal stresses and deformations for the expansion level of 0.30%.

Figure 6.17 shows the evolution of the load and the maximum crack opening at failure along the expansion levels. The maximum crack opening is a value calculated from Eq. (6.1). For panels with an expansion level of 0.04% the maximum crack opening is approximately 1.4 mm and for the panel without expansion the crack opening is approximately 0.77 mm—an increase of 82%. For panels with the maximum expansion level the maximum crack opening is approximately 1.64 mm—an increase of 113% over the maximum panel crack opening without expansion.

$$w = (\varepsilon_t - \varepsilon_{tm}) * leq \tag{6.1}$$

Taking into account all the results presented throughout this section, it can be concluded that the effects of reductions in the mechanical properties of concrete are relevant in the loss of load capacity, in the increase of damage in stiffness, in the increase of crack propagation and in the maximum crack opening. The results show that the above-mentioned effects were significant even at small expansion levels. This indicates the importance of preventive measures in the early stages of expansion.

6.3 Panel S2-1 Deteriorated by Delayed Ettringite Formation (DEF)

In this section results from numerical analyses of Sankovich's panel deteriorated by DEF expansions is presented and discussed. Figure 6.18 shows the evolution of the first cracking and failure loads with the expansion levels, for the fully affected panel. It can be seen that the intact panel, represented by the point of no expansion, presented a first cracking load of about 421 kN and 325.1 kN, for the low expansion level. This means that for a relatively low level of expansion (0.11%), the first cracking load decrease was approximately 23%. For the high expansion level (1%), the first cracking load was 248.6 kN—a decrease of 41%. This fact is consistent

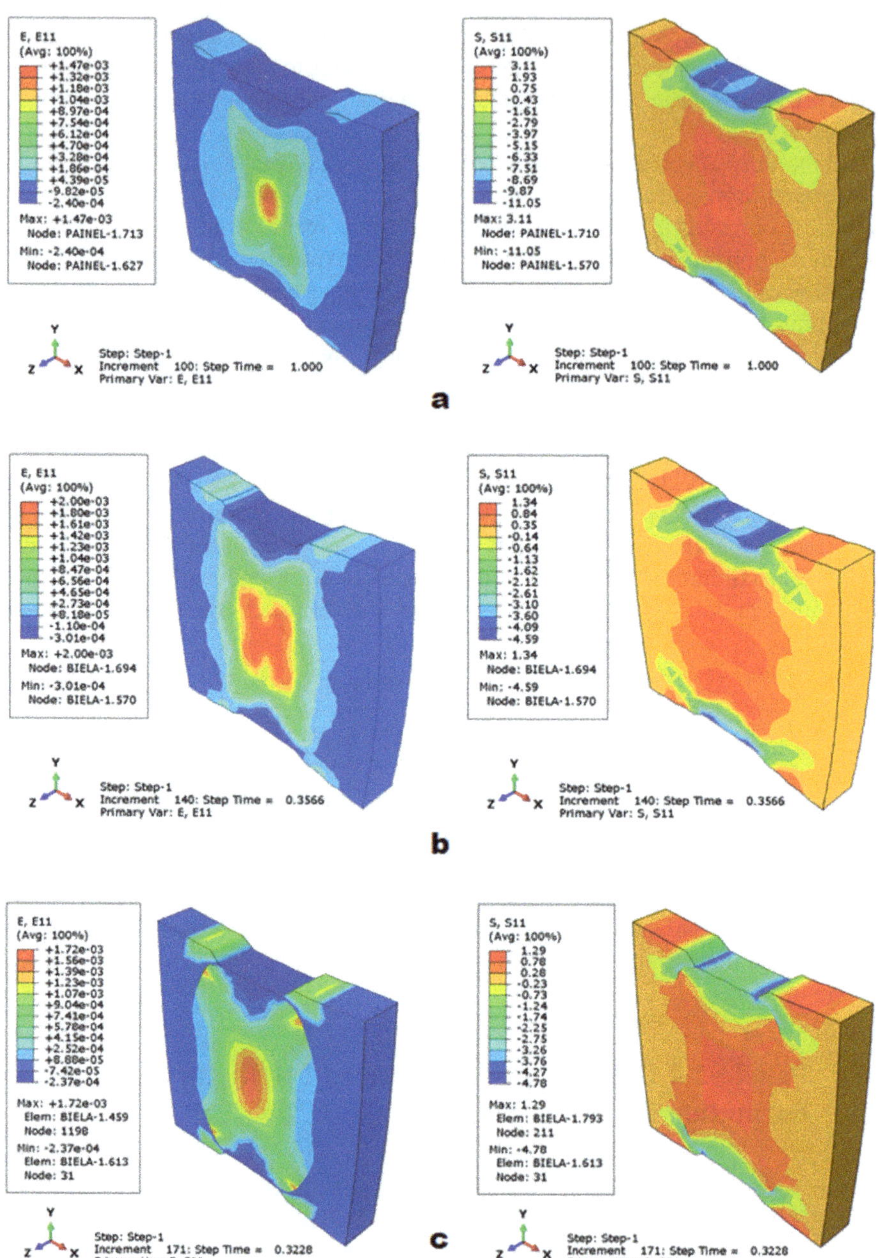

Fig. 6.16 Total deformation and x-axis voltage for expansion of 0.30%. **a** Unaffected. **b** Totally affected. **c** Partially affected

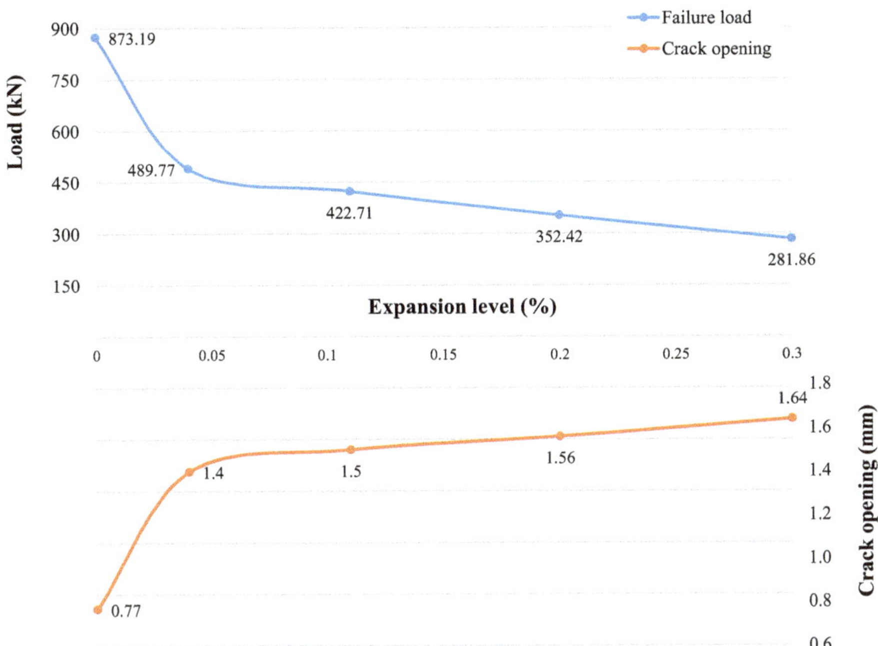

Fig. 6.17 Load and crack opening at failure for each expansion level

with previous researches, which observed that an expansion of 0.1% corresponds to the first appearance of visible cracks in concrete elements affected by DEF. The decrease in the load capacity of the panel observed in a macroscopic level can, in fact, be microscopically explained by the separation between paste and aggregate, as commonly reported by several researchers (Brunetaud et al. 2007; Lin 2019).

The failure load of the intact panel was 706.8 kN. For the fully affected panel in expansion level of 0.11% the failure load reached 490.9 kN—a decrease of 31%. The failure load for the 1% expansion level was even lower—323.4 kN—, representing a decrease of approximately 54% compared with the intact panel. These results highlight the important effect of internal expansion reactions on overall panel behaviour. Similar results were found by Karthik et al. (2020), that reported the energy absorption capacity of large reinforced concrete beam-column joints subject to ASR/DEF deterioration close to 40% of that of control specimen without deterioration.

It is also possible to observe in Fig. 6.18 that there are important decreases in the initial levels of expansion, both at the first cracking load and at failure load. Additionally, it was observed that the effect of the discrete expansions was more pronounced at failure load. For medium expansion levels, the decrease in the first cracking and failure load were less pronounced. This fact points out to the importance of early diagnosis of the levels of expansion existing in concrete elements affected

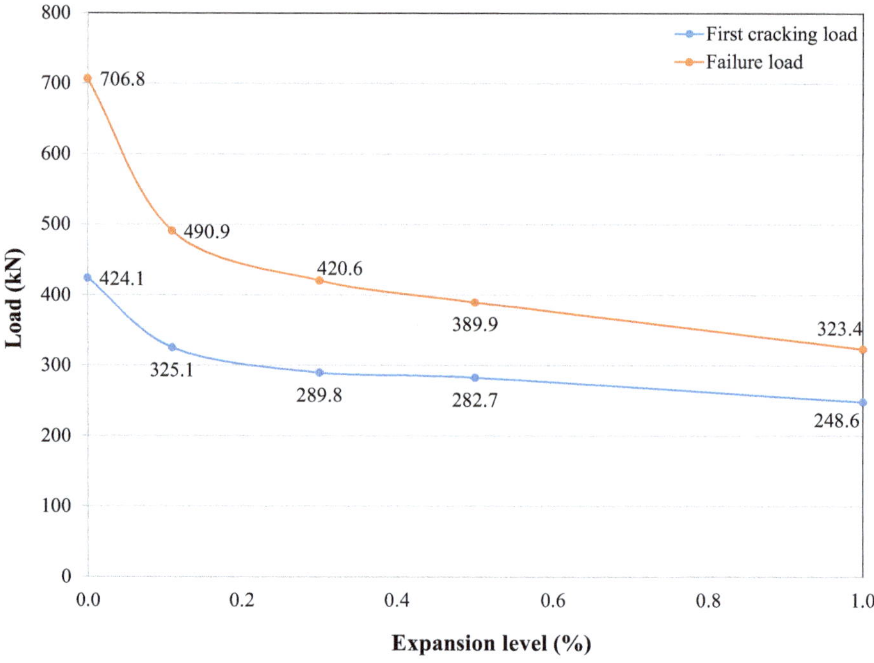

Fig. 6.18 First cracking and failure loads for each expansion level for the fully affected panel

by DEF, so that one can decide regarding the need for recovery or rehabilitation interventions works.

Same analyses were performed for the partially affected panel and the results are shown in Fig. 6.19. It can be observed decreases of 53 and 78% in the first cracking load for low and high levels of expansions, respectively. In the same way, it was observed decreases in the failure load of 25 and 45%, for those levels of expansion.

Figure 6.20 shows the evolution of the maximum crack opening at failure with the expansion levels for both fully and partially affected panel. One can observe that, for low level of expansion, the maximum crack opening is approximately 3.65 mm for the fully affected panel while for the intact panels, the maximum crack opening is approximately 1.76 mm—an increase of 104%. For an expansion level of 1%, the maximum crack opening is approximately 4.51 mm for the fully affected panel—an increase of 156%.

It can also be seen in Fig. 6.20 the evolution of the maximum crack opening at failure with the expansion levels for the partially affected panel, from which one can realize increasing in cracks opening of 208% and 281%, for low and high level of expansions, respectively. Values of crack openings obtained might be compared by order of magnitude with those found in concrete beams deteriorated due to DEF expansion by others researches (Karthik et al. 2016).

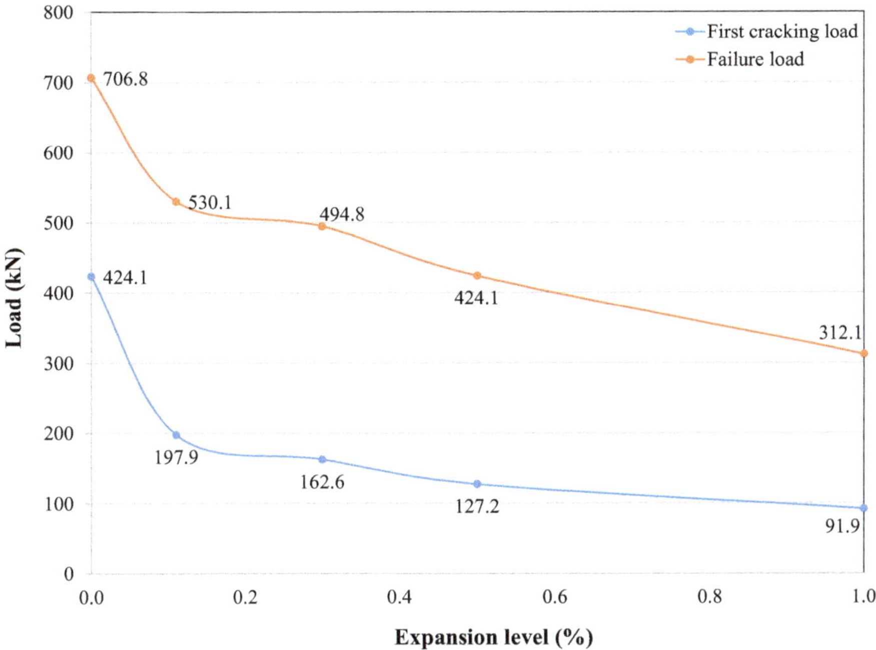

Fig. 6.19 First cracking and failure loads for each expansion level for the partially affected panel

The increase in the crack opening demonstrates the potential damage of DEF expansions in concrete, since these cracks open ways to the transport of aggressive agents from outside to the inside of the element, compromising its durability. Consistently with what was observed, Karthik et al. (2016) states, additionally, that DEF induced cracks can lead to moisture ingress and rebar corrosion which can negatively impact the strength and ductility of the structure.

Compressive strain and damage evolution with the level of expansion at failure for the fully affected panel are shown in Fig. 6.21. It can be seen that the intact panel presented a compressive strain of about 4.7 ‰. For a level of expansion of 0.11%, the compressive strain was about 5.1 ‰ and for the expansion level of 1%, it reached to a compressive strain of 7.4 ‰. This means that, for low level of expansion, the increase in compressive strain was approximately 8.5%, while that, for the high level, the increase was 57.4%. Also, when one looks at the compressive damage, the intact panel exhibited a value of approximately 71%. For the high level of expansion, the damage of approximately 98%—an increase of 38%. These results show a serious decrease in the panel stiffness caused by the DEF expansions.

Compressive strain and damage evolution with the level of expansion at failure for the partially affected panel are shown in Fig. 6.22. It was observed increases of about 31% in compressive strain for low level of expansion and virtually no differences for high level of expansion.

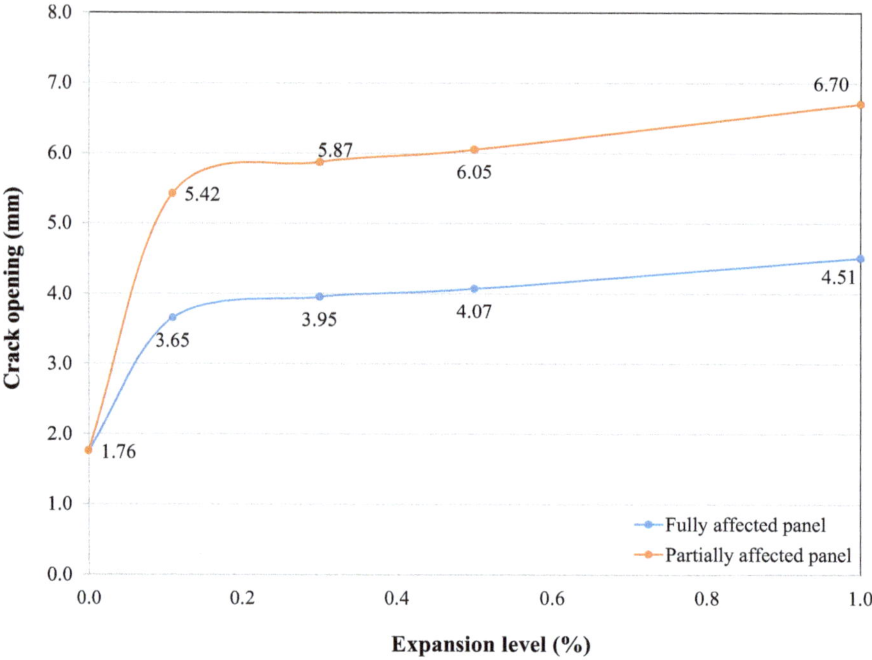

Fig. 6.20 Crack opening at failure for each expansion level for both fully and partially affected panel

Figure 6.23 shows the tensile damage profile at failure for a level of expansion of 1% for the intact, fully affected and partially affected scenarios. The tensile damage profile of the panels is similar, with cracks concentrating in the centre of the panels. It can be seen that the partially affected panel, besides presenting the tensile damage distributed throughout the struts, exhibited damage about 30% on the panel's side and about 50% close to the loading area and the nodal region. The fully affected panel showed damage neither on the panel nor on the nodal region. These results show that in the scenario with the swelling expansion concentrated only in the strut region produced the worst structural performance for the panel investigated. This is an important conclusion for structural design engineers.

Figure 6.24 illustrates the compressive damage profile at failure at the high level of expansion for the intact, fully affected, and partially affected panels. The compressive damage profile of the intact and fully affected panel is quite similar. The concrete crushing is concentrated at the beginning of the compression cone. In the partially affected panel, the compressive damage is distributed throughout the compression cone and the compression wedge worsens, as a result of flattening on the loading

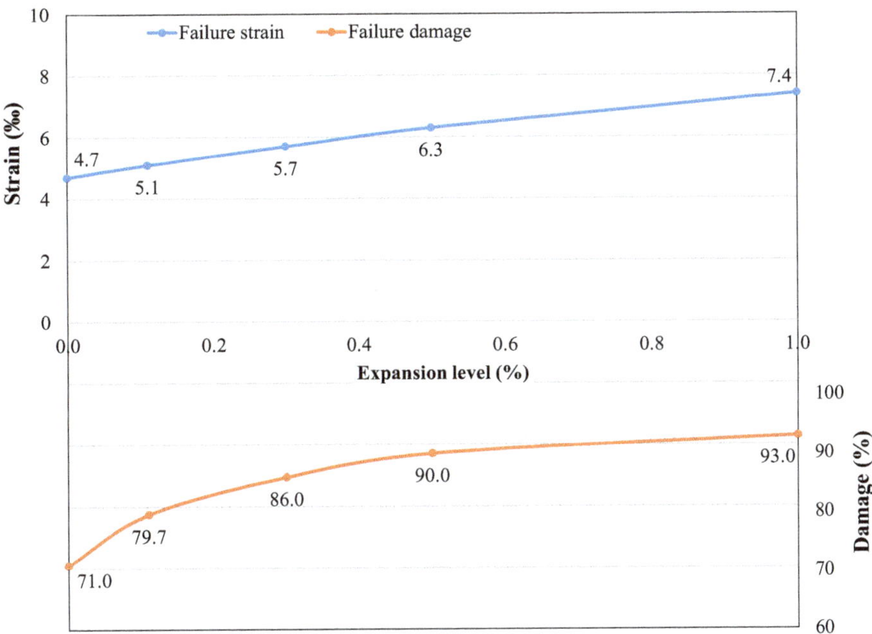

Fig. 6.21 Compressive strain and compressive damage for each expansion level for the fully affected panel

surface. This represents a loss of stiffness at the load transfer points. Compressive damage becomes more evenly distributed in the partially affected panel, which indicates more regions being crushed, even though the value of the damage in this scenario was not the biggest one. This may be happened because the failure of the partially affected panel is governed by cracking in tension.

Taking into account all the results presented this section, it can be concluded that the effects of the in decreases the mechanical properties of the concrete due to DEF expansions are relevant in the loss of load capacity, stiffness and serviceability condition of the concrete panels investigated. Table 6.2 summarizes results discussed in this section.

It is important to highlight that internal swelling reactions induce free expansions in plain concrete as those discussed in the paper. On the other hand, if one deals with reinforced concrete elements, the generated expansions are restrained by the existing reinforcement and, in such cases, reinforcement is stressed in tension while the concrete is stressed in compression. This phenomenon is often referred to as chemical prestressing and its occurrence plays an important role in reinforced concrete elements behavior and cannot be neglected under penalty of loss of relevant

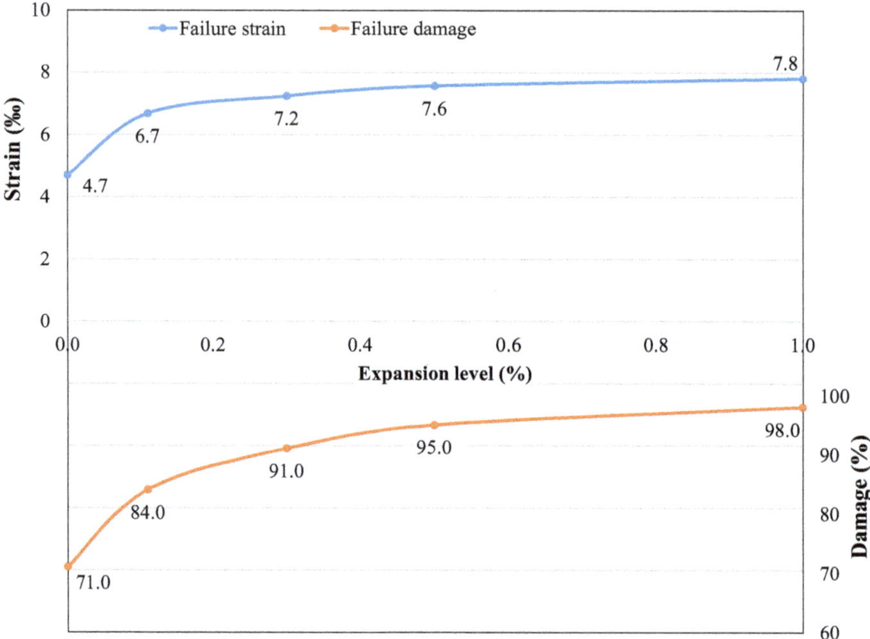

Fig. 6.22 Compressive strain and compressive damage for each expansion level for the partially affected panel

information about the failure mechanism modification, which has a strong influence over the reinforced concrete element stiffness. Some previous researches have already reported that the chemical pre-stress can exert a very significant influence on the load-bearing mechanism of reinforced concrete elements both in service and ultimate limit states (Morenon et al. 2019; Silva 2007). Further investigations are being developed to evaluate the concrete damage models studied to numerically describe such phenomena.

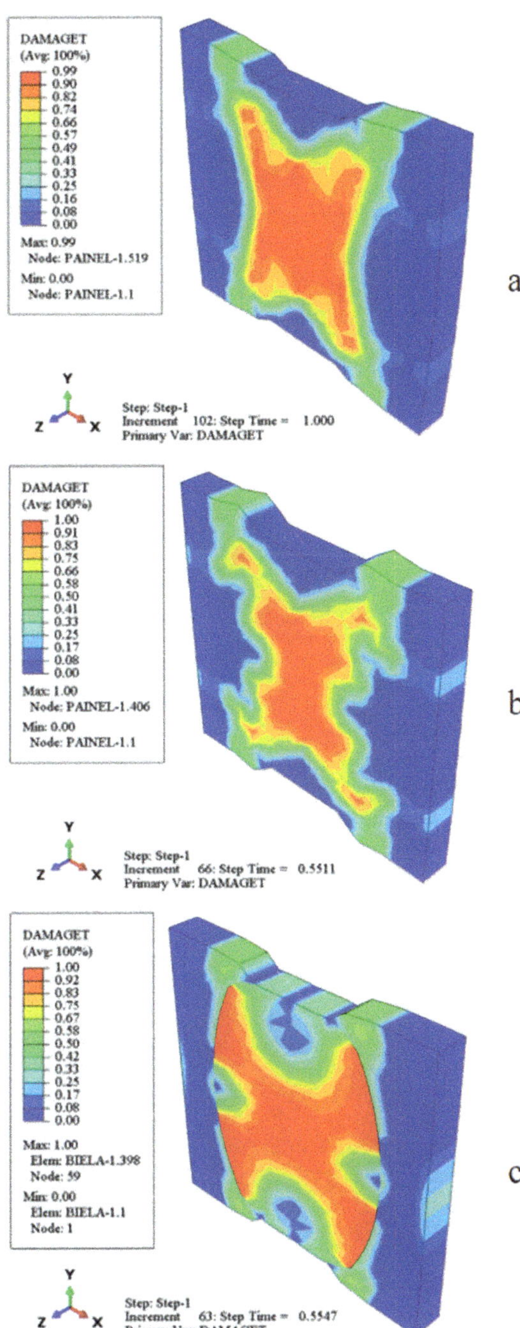

Fig. 6.23 Tensile damage for expansion of 1%. **a** Intact. **b** Fully affected. **c** Partially affected

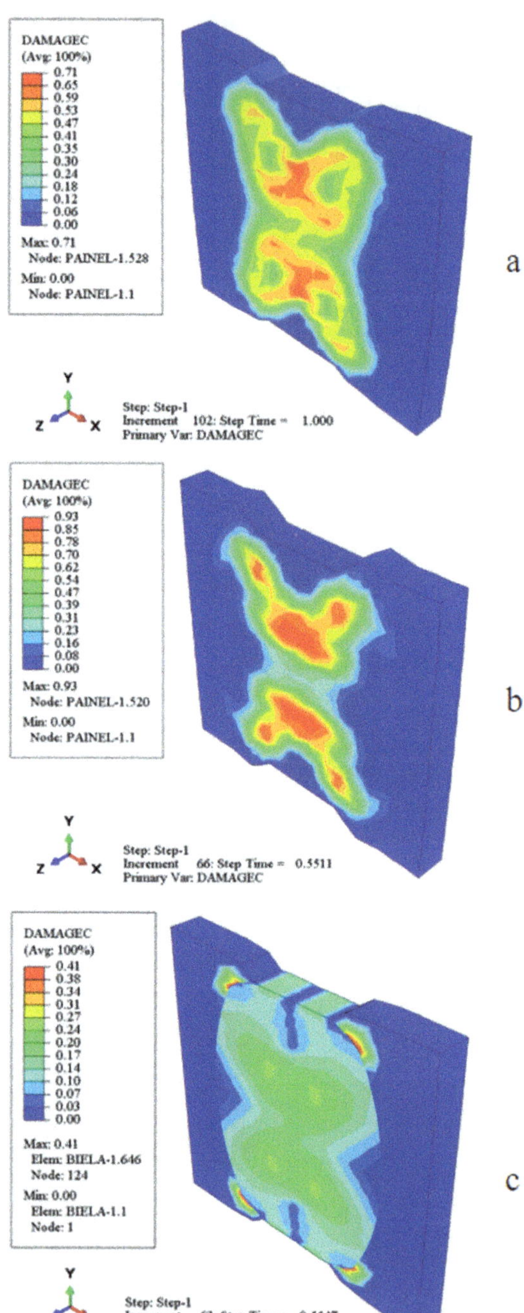

Fig. 6.24 Compressive damage for expansion of 1%. **a** Intact **b** Fully affected. **c** Partially affected

Table 6.2 Summary of decreasing of first cracking and failure loads of panel S2-1

Scenario	Decrease in first cracking load		Decrease in failure load		Increase in crack open width	
	LLE (%)	HLE (%)	LLE (%)	HLE (%)	LLE (%)	HLE (%)
FAP	−23	−41	−31	−54	+104	+156
PAP	−53	−78	−25	−56	+208	+281

FAP = fully affected panel, PAP = partially affected panel, LLE = Low level of expansion, HLE = High level of expansion

References

Brown MD, Sankovich CL, Bayrak O, Jirsa JO, Breen JE, Wood SL (2006) Design for shear in reinforced concrete using strut-and-tie models. Report no. FHWA/TX-06/0-4371-2, University of Texas at Austin

Brunetaud X, Linder R, Divet L, Duragrin D, Damidot D (2007) Effect of curing conditions and concrete mix design on the expansion generated by delayed Ettringite formation. Mater Struct 40(6):567–578

Karthik MM, Mander JB, Hurlebaus S (2016) Deterioration data of a large-scale reinforced concrete specimen with severe ASR/DEF deterioration. Constr Build Mater 124:20–30

Karthik MM, Mander JB, Hurlebaus S (2020) Simulating behaviour of large reinforced concrete beam-column joints subject to ASR/DEF deterioration and influence of corrosion. Eng Struct 222:111064

Lin N (2019) Influence of ASR degradation on structural behavior of concrete structures. MSc thesis, Delft University of Technology, The Netherlands

Morenon P, Multon S, Sellier A, Grimal E, Hamon F, Kolmayer P (2019) Flexural performance of reinforced concrete beams damaged by Alkali-Silica Reaction. Cem Concr Comp 104:103412

Sankovich CL (2003) An explanation of the behavior of bottle-shaped struts using stress fields. MSc thesis, University of Texas, Austin, USA

Silva GA (2007) Retrofitting works of pile caps foundations affected by alkali aggregate reaction. MSc thesis, Catholic University of Pernambuco, Brazil

Chapter 7
Conclusions and Future Recommendations

7.1 Conclusions

The objective of this work was to understand the structural behaviour of isolated bottle-shaped struts deteriorated by internal expansion reactions through numerical modelling with the FEM model. In general, the results obtained showed that the CDPM is a model of plasticity with damage with the potential to represent the load-bearing mechanisms of isolated concrete bottle-shaped struts a range of several stress levels to which these elements may be subjected in the panels investigated. The numerical simulations for various levels of expansion were able to consistently represent the expected damage profile in the panels as well as the estimate of the first crack loading the crack opening and the value of the maximum possible force to be applied. For the panels investigated, the reduction observed in the failure load reached a value of 70%, the increase of the tensile plastic deformation was more than 60% and the maximum crack opening can reach an increase of 113%, when compared with those observed experimentally in panels without internal swelling reactions.

In resume, the main conclusions that can be drawn from the research carried out are:

- The ABAQUS CDPM proved to be an efficient numerical strategy for the representation of mechanical phenomena involved in load support mechanisms in partially loaded concrete specimens with the formation of isolated bottle-shaped struts. Calibration of the CDPM model with Sankovich's experiments (Sankovich 2003) enhanced the capabilities of the model;
- For the use of CDPM, there is, however, an important difficulty to be overcome—the unavailability of experimental results for a varied concrete property range regarding the plastic strains and damage, both in tension and compression regimes;

A. C. Azevedo et al., *Concrete Structures Deteriorated by Delayed Ettringite Formation and Alkali-Silica Reactions*, Building Pathology and Rehabilitation 24,
https://doi.org/10.1007/978-3-031-12267-5_7

- To overcome the difficulties pointed out, a MATLAB script was developed, based on Alfarah's research (Alfarah 2017) that allow obtaining the properties necessary for the use of CDPM in an automated way and with few input parameters—the characteristic compressive strength of the concrete, the equivalent length of the finite element mesh and the ratio between plastic and inelastic strains in compression;
- To perform numerical simulations of concrete with the use of CDPM it is desirable to use a structured and uniform finite element mesh. This is important because the model is relatively sensitive to the equivalent length of the finite element mesh and thus in problems with distinct mesh density will require plastic strains and damages both in tension and in compression regimes, for each region of the model with different equivalent length.

A strategy of numerically simulating the effects of internal expansion reactions on concrete specimens partially considering the effect of these expansions on the properties of concrete strength and strain, according to Sanchez (Sanchez et al. 2017, 2018) was investigated in the research and the following results summarize the main findings:

- Small to moderate internal expansion levels exhibited an important influence on the expected damage profile, the estimation of the first crack load, the estimated crack opening and the load capacity of the investigated specimens, both for ASR expansions;
- For the expansions due to ASR, when compared with the values of the intact specimen, the following findings stand out:
  - The reduction in the load capacity of the specimens studied was 70%;
  - The increase in plastic tensile strain at rupture was 60%;
  - The maximum crack opening increased about 113%.
- For the expansions due to the DEF, when compared with the values of the healthy panel, the following findings stand out:
  - The reduction in load capacity of the panels studied was 78%;
  - The increase in plastic deformation in compression has been doubled;
  - The maximum crack opening experienced an increase of 281%.

7.2 Future Recommendations

Regarding future studies within the area, the authors suggest the following approaches to the theme:

- Perform a sensitivity analysis of viscosity and friction angle parameters that integrate the fundamentals of the CDP rupture surface;

- Numerically study Sankovich models equipped with reinforcements with varied distribution configurations to formulate the understanding of their influence on panels affected by internal expansion reactions;
- Improve the routine developed in such a way that eliminates the dependence on the results of the finite element mesh used.

Acknowledgements This work was supported by: Base Funding—UIDB/04708/2020 and Programmatic Funding—UIDP/04708/2020 of the CONSTRUCT—Instituto de I&D em Estruturas e Construções—funded by national funds through the FCT/MCTES (PIDDAC) and by the Brazilian Research Council—Conselho Nacional de Desenvolvimento Científico e Tecnológico (CNPq).

References

Alfarah B, López-Almansa F, Oller S (2017) New methodology for calculating damage variables evolution in plastic damage model for RC structures. Eng Struct 132:70–86

Sanchez LFM, Fournier B, Jolin M, Mitchell D, Bastien J (2017) Overall assessment of Alkali-Aggregate Reaction (AAR) in concretes presenting different strengths and incorporating a wide range of reactive aggregate types and natures. Cem Concr Res 93:17–31

Sanchez LF, Drimalas T, Fournier B, Mitchell D, Bastien J (2018) Comprehensive damage assessment in concrete affected by different internal swelling reaction (ISR) mechanisms. Cem Concr Res 107:284–303

Sankovich CL (2003) An explanation of the behavior of bottle-shaped struts using stress fields, University of Texas, Austin, USA

Uncited References

Alnaggar M, Luzio GD, Cusatis G (2017) Modeling time-dependent behavior of concrete affected by alkali silica reaction in variable environmental conditions. Materials 10(5):471

Balachandran C, Muñoz JF, Arnold T (2017) Characterization of alkali silica reaction gels using Raman spectroscopy. Cem Concr Res 92:66–74

Bangert F, Kuhl D, Meschke G (2004) Chemo-hygro-mechanical modelling and numerical simulation of concrete deterioration caused by alkali-silica reaction. Int J Numer Anal Methods Geomech 28(78):689–714

Bazant ZP, Steffens A (2000) Mathematical model for kinetics of alkali–silica reaction in concrete. Cem Concr Res 30:419–428

Bazant ZP, Zi G, Mayer C (2000) Fracture mechanics of ASR in concretes with waste glass particles of different sizes. J Eng Mech 126(3):226–232

Benmore CJ, Monteiro PJM (2010) The structure of alkali silicate gel by total scattering methods. Cem Concr Res 40(6):892–897

Bourdot A, Thiéry V, Bulteel D, Hammerschlag JG (2016) Effect of burnt oil shale on ASR expansions: a petrographic study of concretes based on reactive aggregates. Constr Build Mater 112:556–569

Brunetaud X, Divet L, Damidot D (2008) Impact of unrestrained delayed ettringite formation-induced expansion on concrete mechanical properties. Cem Concr Res 38(11):1343–1348

Comi C, Fedele R, Perego U (2009) A chemo-thermo-damage model for the analysis of concrete dams affected by alkali–silica reaction. Mech Mater 41(3):210–230

Collepardi M (2003) A state-of-art review on delayed ettringite attack on concrete. Cem Concr Compos 25:401–407

Dähn R, Arakcheeva A, Schaub Ph, Pattison P, Chapuis G, Grolimund D, Wieland E, Leemann A (2016) Application of micro X-ray diffraction to investigate the reaction products formed by the alkali–silica reaction in concrete structures. Cem Concr Res 79:49–56

Diab SH, Soliman AM, Nokken MR (2020) Changes in mechanical properties and durability indices of concrete undergoing ASR expansion. Constr Build Mater 251:118951

Dunant CF, Scrivener KL (2010) Micro-mechanical modelling of alkali-silica–reaction-induced degradation using the AMIE framework. Cem Concr Res 40(4):517–525

Dunant C (2009) Experimental and modelling study of the alkali-silica-reaction in concrete. PhD thesis, Lausanne, EPFL, Switzerland

Ghanem H, Zollinger D, Lytton R (2010) Predicting ASR aggregate reactivity in terms of its activation energy. Constr Build Mater 24(7):1101–1108

Grimal E, Sellier A, Le Pape Y, Bourdarot E (2008a) Creep, shrinkage, and anisotropic damage in AAR swelling mechanism—Part I: A constitutive model. ACI Mater J 105(3):227–235

Grimal E, Sellier A, Le Pape Y, Bourdarot E (2008b) Creep, shrinkage and anisotropic damage in AAR swelling mechanism—Part II: Identification of model parameters and application. ACI Mater J 105(3):236–242

Hobbs DW (1988) Alkali–silica reaction in concrete. Thomas Telford, London

Jiang Z, He B, Zhu X, Ren Q, Zhang Y (2020) State-of-the-art review on properties evolution and deterioration mechanism of concrete at cryogenic temperature. Constr Build Mater 257:119456

Lindgård J, Andiç-Çakir Ö, Fernandes I, Rønning TF, Thomas MDA (2012) Alkali–silica reactions (ASR): literature review on parameters influencing laboratory performance testing. Cem Concr Res 42(2):223–243

Lin N (2019) Influence of ASR degradation on structural behavior of concrete structures. MSc thesis, Delft University of Technology, The Netherlands

MATLAB (2012) The language of technical computing. The MathWorks, Inc. http://www.mathworks.com

MERL Report-09-23, New recommendations for ASR mitigation in reclamation concrete construction. U.S. Bureau of Reclamation, Denver, Colorado, USA

Merz C, Leemann A (2013) Assessment of the residual expansion potential of concrete from structures damaged by AAR. Cem Concr Res 52:182–189

Multon S, Toutlemonde F (2010) Effect of moisture conditions and transfers on alkali silica reaction damaged structures. Cem Concr Res 40:924–934

Multon S, Sellier A, Martin C (2009) Chemo-mechanical modeling for prediction of alkali silica reaction (ASR) expansion. Cem Concr Res 39:490–500

Murdoch R (2015) Creating a simplified model of the alkali-silica reaction in concrete by utilizing finite element modelling techniques. MSc thesis, University of Southern Queensland, Australia

Pignatelli R (2012) Modeling of degradation induced by alkali-silica reaction in concrete structures. PhD thesis, POLIMI, Italy

Rajabipour F, Giannini E, Dunant C, Ideker JH, Thomas MDA (2015) Alkali–silica reaction: current understanding of the reaction mechanisms and the knowledge gaps. Cem Concr Res 76:130–146

Sahoo DK, Gautam RK, Singh B, Bhargava P (2008) Strength and deformation of bottle-shaped struts. Magaz Concr Res 60(2):137–144

Steffens A, Li K, Coussy O (2003) Ageing approach to water effect on alkali-silica reaction. Degradation of structures. J Eng Mech 129:50–59

Swamy RN (1992) The alkali-silica reaction in concrete. Taylor-Francis, CRC Press

Thomas MDA, Fournier B, Folliard KJ, Resendez YA (2011) Alkali-silica reactivity field identification handbook, No. FHWA-HIF-12-022, US Department of Transportation, USA

Ulm FJ, Coussy O, Kefei L, Larive C (2000) Thermo-chemo-mechanics of ASR expansion in concrete structures. J Eng Mech 126:233–242

GPSR Compliance
The European Union's (EU) General Product Safety Regulation (GPSR) is a set of rules that requires consumer products to be safe and our obligations to ensure this.

If you have any concerns about our products, you can contact us on

ProductSafety@springernature.com

In case Publisher is established outside the EU, the EU authorized representative is:

Springer Nature Customer Service Center GmbH
Europaplatz 3
69115 Heidelberg, Germany

www.ingramcontent.com/pod-product-compliance
Ingram Content Group UK Ltd.
Pitfield, Milton Keynes, MK11 3LW, UK
UKHW021834270726
14058UKWH00001B/151
9783031122699